Routine Cytological Staining Techniques

Routine Cytological Staining Techniques

Theoretical Background and Practice

Mathilde E. Boon

and

Johanna S. Drijver

MACMILLAN

First published 1986

Published by
MACMILLAN EDUCATION LTD
Houndmills, Basingstoke, Hampshire RG21 2XS
and London
Companies and representatives
throughout the world

Filmsetting by Vantage Photosetting Co. Ltd
Eastleigh and London
Printed in Hong Kong

British Library Cataloguing in Publication Data
Boon, Mathilde
Routine cytological staining techniques.
1. Stains and staining (Microscopy)
2. Cytology—Technique
I. Title   II. Drijver, Johanna
578′.64   QH237
ISBN 0–333–38316–8
ISBN 0–333–39713–4 Pbk

Contents

PART TWO: Atlas

Preface

We had several reasons for writing this book and several goals in mind when we started our investigations into routine fixation and staining. Above all we wanted to know more about the mechanisms of the staining processes, and to gain an insight into the differences between staining cytological and histological material. We believe that the results of our research are of considerable assistance to anyone involved in routine cytological diagnosis, since the quality of the diagnosis depends on the quality of the smear. We do not consider immunotechniques and enzyme staining to be routine, so these methods are not discussed in our book.

We have learned that when the mechanisms of the staining processes are understood, one can manipulate the staining results according to one's wishes, and standardise the staining. Both goals are most easily reached if the staining solutions are prepared in the laboratory, a process that has proved to be easier than we had anticipated. This manual therefore contains details of methods for preparing staining and fixing solutions, together with practical advice on the most economical methods of staining and fixing.

We also discuss the interpretation of the staining results, and the need to be critical of unnecessary sub-optimal results. We stress that cellular and nuclear dimensions in cytological slides are totally dependent on the cytopreparatory technique employed, and differ significantly from those in tissue sections.

The idea that the quality of the slide is of paramount importance for the success of cytodiagnosis was first realised by one of the authors (M.E.B.) when she was trained in the laboratory of Dr B. Naylor in Ann Arbor, Michigan, during the period 1967–68. In the following years, we noticed at various cytology congresses that many cytological diagnoses were made on sub-optimal slides, and the cytopathologists did not even know exactly how these were fixed and stained. In discussions we became aware that neither we nor our fellow cytologists knew exactly what component in the routine stain stained what in our slides. The explanations we found in cytology and histology books concerning staining cytological slides were not satisfactory for us. We therefore realised that we had to investigate for ourselves in order to find the right answers. Since we thought that knowledge concerning the components of the cell was a keystone in such investigations, we sought contact with the world of biology, and therefore this book is written by a cytologist and a biologist. The very satisfying contacts with other research workers in Europe, Japan and the USA have greatly influenced our insight into this complicated subject. When we had acquired a knowledge of staining mechanisms, we were able to exploit the staining techniques instead of suffering by them, and to profit from their use in our diagnostic work.

It should be clearly understood that our work is far from complete. We hope, however, that this book will enable readers to find the preparation and staining of cytological slides to be as exciting and satisfying as the authors do themselves.

Leiden, 1985

M.E.B.
J.S.D.

Acknowledgements

Many people have given us invaluable advice concerning this book. We thank them all, in particular the following.

Dr R.W. Horobin of the University of Sheffield, UK, read through the first draft of the book and gave us priceless constructive criticism. He also provided us with the data on the physical and chemical characteristics of the dyes, as published in this book. Professor Dr D. Wittekind of the University of Freiburg, West Germany, has shown from the beginning of our work a great interest in our book, and has shared with us the results of his own experiments with the Romanowsky–Giemsa staining method.

We are grateful for the beautiful scanning electron microscope pictures, which form an essential part of the book by showing the three-dimensional aspects of cells in smears. They were provided by Ms P.H.T. van der Zanden and Dr P. Kenemans from the division of gynaecological oncology, University of Nijmegen, The Netherlands.

Last but not least we want to thank the cytotechnologists of the Leiden Cytology Laboratory for their endurance when trying to do cytodiagnostic work on smears stained in all colours of the rainbow, and the cytotechnologists of the SSDZ in Delft for their co-operation in developing various cytopreparatory techniques.

The drawings in this book were made with great enthusiasm by B. Brest from the Technical University in Delft. The cover is based on the colours of stained cells in cytological slides. Our good friend Frans Rombouts, the artist who designed it, succeeded extremely well in demonstrating how beautiful staining can be.

Part One

1

Cell Biology

In order to know how the different reagents used in cytopreparatory techniques react with the two main components of the cell (the nucleus and the cytoplasm), it is necessary to know their respective compositions.

1.1 Nucleus

In its interphase the nucleus shows a more or less diffuse granular structure interspersed with larger granules. Another feature evident in this phase is the nucleolus, of which there can be one or more appearing as dense globular structures. The nuclear envelope that separates nucleus and cytoplasm is only visible as such by means of an electron microscope.

1.1.1 Chromatin

When not dividing, the nucleus consists of a mass of granular material called chromatin (meaning 'an intensely staining substance'). The more diffuse part of it forms the euchromatin, and its coarser structure is the heterochromatin. Chromatin is the material of which the chromosomes are composed. During the interphase the chromosomes are unrecognisable as separate units because they are in an elongated state, dispersed throughout the nucleus and entwined like pieces of wool. This is in sharp contrast to the condensed, coiled chromosomes that can be seen during cell division. Not all of the chromatin uncoils after cell division. Some of it remains condensed in the form of heterochromatin. One of the X-chromosomes, the so-called Barr body, is always visible in the condensed form during both interphase and when the cell is dividing.

The structure of chromatin is shown in figure 1.1. It can be likened to a row of balls held together by a string of DNA molecules wound around them. The unit 'balls plus string' is known as the nucleosome structure. The 'balls' consist of histones, the characteristic basic proteins of the chromosomes, which are linked to DNA molecules by ionic bonding

(a)　　　　　　　　　　　(b)

Figure 1.1　The structure of the elementary chromatin fibre. (a) Before fixation. (b) After fixation: the DNA double helix after separation of the DNA and histones by means of a fixative.

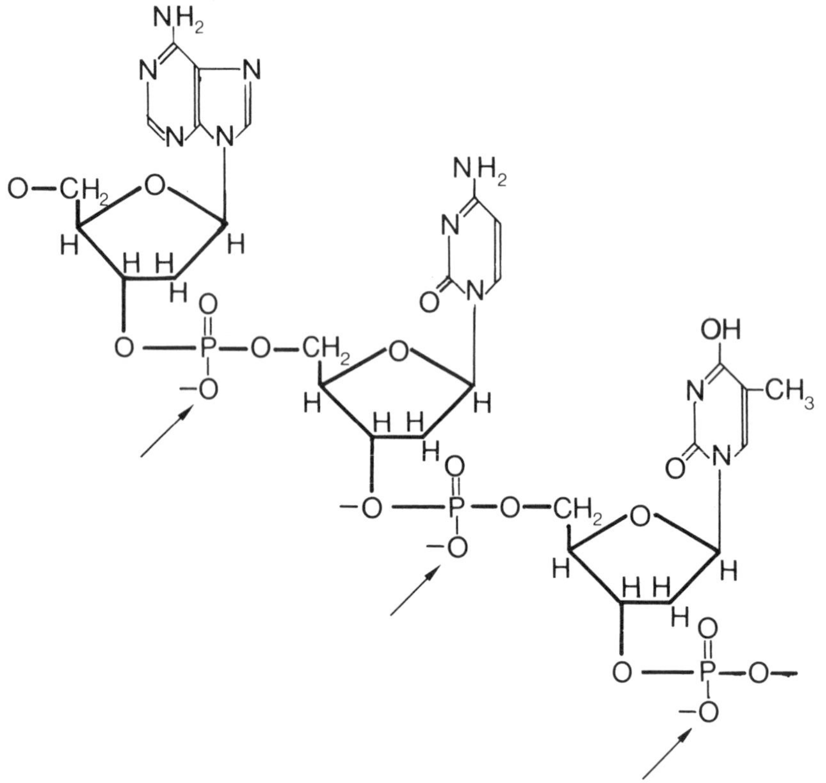

Figure 1.2(a)   Three nucleosides of the DNA molecule linked together by phosphoric acid to form three nucleotides. Arrows indicate sites of negative charge.

between phosphorus atoms and the cationic groups on their own molecules.

Chemically speaking, there is no difference between euchromatin and heterochromatin. It is only their physical structure, specifically the degree of coiling, which is different. In euchromatin the elementary chromosome fibres are coiled only slightly, if at all. In heterochromatin, they are more coiled, and the Barr body is in an extreme state of coiling. There is also, however, a difference in their activity: transcription between the chromosomes does not occur when they are in their condensed form but when they are in their elongated state. In the interphase nucleus, only those DNA strands needed for the activities of the cells are uncoiled. Undifferentiated cells, such as stem cells in bone marrow, have a diffuse chromatin, whereas normoblasts, for example, have a compact chromatin

pattern because of the presence of a larger amount of heterochromatin.

1.1.2 Composition and Structure of DNA and RNA

The nucleic acids are polymers of small units called nucleotides. Each nucleotide results from linking a pentose (either deoxyribose or ribose) and an organic base to form a nucleoside, these units in turn being linked together by phosphoric acid. Structurally speaking, the backbone of the nucleic acid molecule is formed by the pentose and the phosphoric acid, with the organic bases attached to the pentose ring (see figures 1.2a and b) forming a long nucleotide strand. This structure uses two of the three bonding sites which are available on the phosphoric acid molecule, leaving one site for

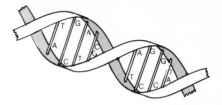

Figure 1.3 The double helix of DNA. Two strands of nucleotides form a double spiral.

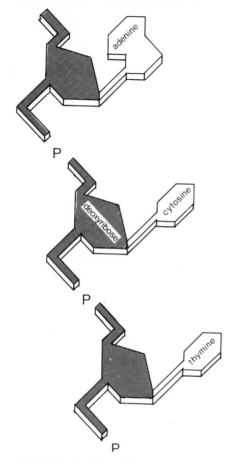

Figure 1.2(b) The backbone of the DNA molecule and its 'spikes' of organic bases.

linking with the histone of the nucleosomes to form chromatin (see figure 1.1).

1.1.2.1 DNA

The DNA molecule is made up of two nucleotide strands linked together by hydrogen bonding, with the organic bases facing each other. This double strand of polymers is twisted in the form of a helix and then spiralled again to form a double helix (figure 1.3). The organic bases on each strand of the molecule are always linked with the same partner on the other strand, but the sequence in which the pairs occur on the backbone is specific and forms the code of information for protein synthesis. The information thus encoded in the DNA molecule is passed on to the cell's cytoplasm via RNA. The mechanism of information transfer can best be studied in more detailed cell biology books. Here it is sufficient to know that the amount of DNA in the nucleus is constant for cells in the interphase, but will be different for cells engaged in cell division.

There are four different organic bases in the DNA molecule: the purines adenine (A) and guanine (G), and the pyrimidines cytosine (C) and thymine (T), which occur in the combinations A–T and C–G (see figures 1.3 and 1.4). Under most staining conditions, the DNA molecule will be ionised, with a negative charge in the phosphoric acid group.

The bonding forces that hold together the double helix (hydrogen and hydrophobic bonding) are not very strong and can be broken by heating or by treatment at high pH (> 7.0). Both methods lead to separation of the two strands (denaturation) of the DNA molecule. This process is not to be confused with denaturation of proteins (see section 1.2.1.2). Denaturation of DNA is reversible whereas that of proteins is irreversible.

1.1.2.2 RNA

The pentose in RNA is ribose and the organic bases are uracil (U), adenine (A), cytosine (C) and guanine (G) (see figure 1.4). In RNA, uracil replaces the thymine found in DNA so that the following pairs are formed: A–U and C–G. Apart from their chemical differences there are also differences in the spatial structure between DNA and RNA. The nucleotide strand of RNA remains single but may fold itself up, forming hairpin loops which can give the molecule a very compact structure.

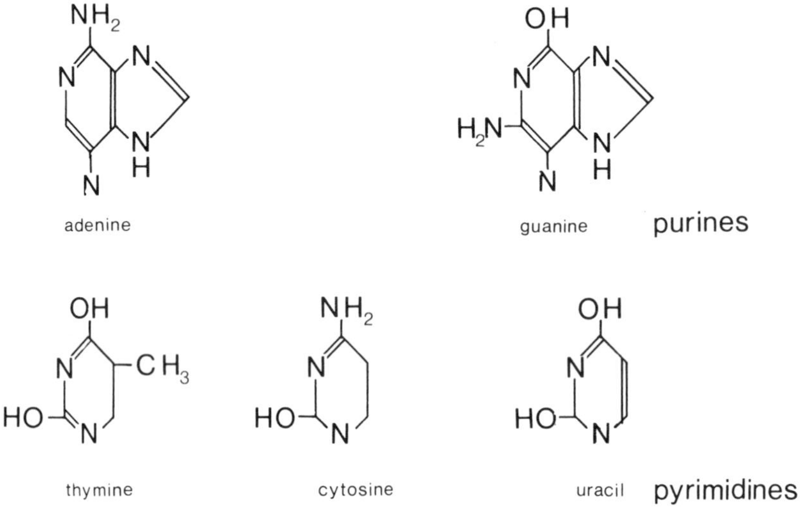

Figure 1.4 **The organic bases of DNA and RNA.**

According to their function there are three kinds of RNA. Messenger RNA (mRNA) transmits the code of information which is transcribed through the sequence of the nucleotides in the RNA molecule to the cytoplasm. This information is used for protein production on the ribosomes. The ribosomes also contain some RNA (rRNA), which is formed in the nucleolus (see section 1.1.4). The third type of RNA is transfer RNA (tRNA), which is responsible for the positioning of the amino acid groups on the polypeptide chains formed in the cytoplasm. The three types of RNA differ in length of molecule and spatial configuration.

In contrast to DNA, the amount of RNA in the nucleus varies with the cell's activity. More RNA will be present with more protein synthesis. Because

Table 1.1   A summary of the chemical composition of the nucleus

Nuclear component	Charged group	Quantity
Nucleic acids		
DNA		fairly constant
	phosphoric acid:	
RNA	$\begin{array}{c} \nearrow O \\ P \\ \searrow O^- \end{array}$	varying, depending on cell's activity
Nucleoprotein		
histones	amino acids: arginine, lysine: NH_3^+	fairly constant
non-histone proteins	amino acids: aspartic, glutamic and hydroglutamic: 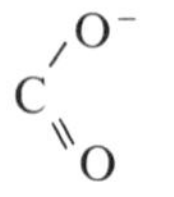 $\begin{array}{c} \nearrow O^- \\ C \\ \searrow O \end{array}$	varying, depending on cell's activity

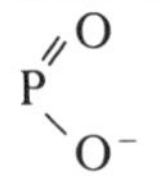

of its role, its presence can be demonstrated in the nucleus, nucleolus and cytoplasm.

1.1.3 Nucleoproteins

Proteins form the main part of the cell, in the nucleus as well as in the cytoplasm. A more detailed description of their structure and nature is given in section 1.2.1. The proteins of the nucleus can be divided into two groups, the histones and the non-histone proteins.

There are five different types of histones. They play an important role in maintaining the spiral structure of the DNA molecule. Their mass is fairly constant and is the same as that of DNA. Because of their high content of the amino acids arginine and lysine, which have positively charged NH_4-groups, they are basic in reaction and are attached in a salt-like union to the nucleic acids.

The non-histone nucleoproteins occur in a very large number of types. Their quantity also varies. Hence, unlike the histones, the ratio of non-histone proteins to DNA varies. Cells with active genes have more non-histone proteins than genetically inactive cells. The non-histone proteins are acid in reaction.

Summarising, one can say that active cells contain more RNA and more non-histone proteins than less active cells. Both RNA and non-histone proteins' are acid so an active cell will have a larger positive charge than a less active cell. The nuclear components and their reactive (charged) groups are summarised in Table 1.1.

1.1.4 Nucleolus

A cell can contain one or more nucleoli. After staining, the nucleolus can be seen under the light microscope as a dense, round particle with roughly round structures inside it. In live cells, round vacuoles can sometimes be seen inside the nucleolus, by means of phase contrast microscopy, corresponding to the roughly round structures seen

when using light microscopy. These are known as nucleolini. By means of phase contrast microscopy, some dense material can be seen around the nucleolini in the live cell. Thus the nucleolus has two zones, less dense (the nucleolini) and dense. Cells which are known to be protein manufacturers (e.g. oocytes and secretory cells) have larger nucleoli than cells that carry out little protein synthesis (e.g. spermatocytes, blastomers and muscle cells).

The nucleoli disappear during cell division and are formed again at telophase. This happens at the nucleolar organising region of the chromosomes.

That RNA is one of the constituents of the nucleolus can be concluded from its staining with Pyronin Y (see section 5.4.3) and its absorption of ultraviolet light at 260 nm, which is also typical. RNA constitutes 3 to 10% of the nucleolar mass and the rest consists of non-histone ribonucleoproteins. Among these are ribosome precursors, which is a clear indication that the nucleolus is the site of ribosome synthesis. Love *et al.* (1973) were able to demonstrate nine different ribonucleoproteins after blocking of the amino groups of the nucleoproteins, and DNA- and RNA-digestion, dyeing with Toluidine Blue and after-treatment with molybdate. We shall describe this technique further in section 7.2.3.

The inner zone of the nucleolus contains some DNA (Feulgen positive). There may be a ring of Feulgen-positive heterochromatin around the nucleolus, some of which may penetrate into it. This is the so-called nucleolus-associated chromatin.

In electron micrographs a clear distinction becomes visible between the two zones seen in phase contrast and light microscopy. There is a fibrillar zone, usually near the centre and corresponding to the nucleolini, and a granular zone, which is more electron dense. Both zones contain RNA. In cells active in protein synthesis, the fibrillar centre contains ribosomal DNA involved in RNA transcription whereas the granular zone contains ribosomal precursors at different stages of completion. The nucleolus can be dyed differentially by the routine staining techniques of Papanicolaou and Romanowsky-Giemsa (see Atlas section, plates 5 and 12); however, factors such as the fixative used and the pH of the staining baths have an important bearing on the resulting colour (see sections 4.2 and 9.3.6). The important role the nucleolus plays in protein synthesis and hence in cell growth has stimulated much research using qualitative and quantitative dyeing techniques. In qualitative studies the colour, shape, number and size of the nucleoli are used in the diagnosis of malignant cells. In quantitative studies either the amount of ribonucleoproteins or the RNA content of the nucleolus is measured. Some of these techniques are discussed in section 5.4.3. For a more detailed study of staining of the nucleolus, see chapter 7.

1.2 Cytoplasm

In this section we shall take a closer look at the cytoplasm of the cell. Protein is its main constituent, but varying amounts of RNA, saccharides and lipids are also present in the cytoplasm, depending on the activity of the particular cell. In addition, secretory products and varying amounts of keratin and its precursors may be present.

RNA has already been dealt with in section 1.1.2.2. Here we shall discuss proteins, saccharides and lipids.

Of the organelles in the cytoplasm (mitochondria, ribosomes and Golgi apparatus) only the mitochondria are sometimes stained (see section 8.2.2).

The reactivity of the chemical constituents of the cytoplasm that are discussed here are summarised in Table 1.2.

1.2.1 Proteins

Proteins are built up of large numbers of sub-units, the amino acids, which are joined together by peptide linkages forming polypeptides. If such a combination of amino acids forms a unit of which the molecular weight exceeds 5000, it is called a protein; if it is smaller than 1500, it is known as a polypeptide.

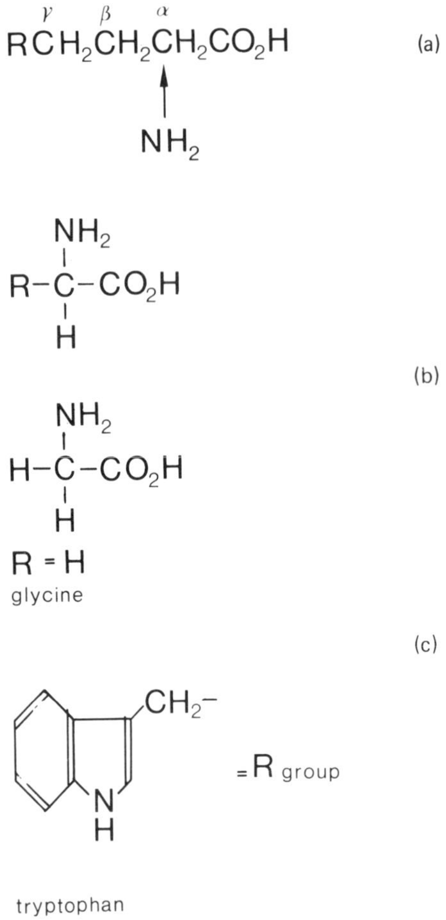

Figure 1.5   (a) Carboxylic acid, showing the position of the α-carbon atom. (b) The general formula of an amino acid. (c) Two amino acids; the simplest, glycine, and a more complicated one, tryptophan.

Amino acids themselves are carboxylic acids to which an NH_2 group is added. In the amino acids found in nature, the amino group is nearly always attached to the α-carbon atom (see figure 1.5a). Thus the general formula for amino acids is as shown in figure 1.5b. The R group can be simply H, as in glycine, or much more complex as, for instance, in tryptophan (see figure 1.5c).

Most amino acids possess at least two ionisable groups, namely the NH_2 group and the COOH group, to which the R group may add another ionisable group. If the R group does not possess an

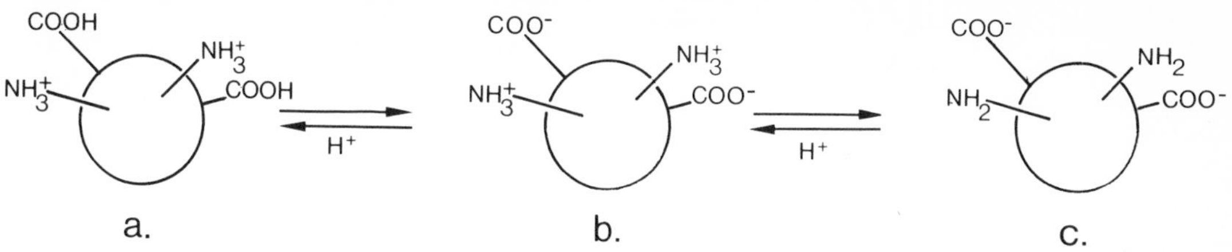

a. b. c.

Figure 1.6 Charges on a protein molecule (a) in acid solution; (b) at IEP; (c) in alkaline solution. (After DeRobertis and DeRobertis, 1980).

$$\underset{R_1}{H_2NCHCO_2H} \; + \; \underset{R_2}{H_2NCHCO_2H} \longrightarrow \underset{R_1 \quad R_2}{H_2NCHCONHCHCO_2H} \; + \; H_2O$$

Figure 1.7 The peptide linkage.

ionisable group, the amino acid is neutral. If it possesses an additional NH_2 group, the amino acid becomes basic, and it becomes acid if there is a COOH group present. The majority of naturally occurring amino acids are neutral. The acidic amino acids aspartic acid and glutamic acid can be found in the non-histone proteins, whereas the basic amino acids lysine and arginine are part of the histone proteins. The charges on a protein molecule are shown in figure 1.6.

1.2.1.1 Primary Structure

The peptide linkage between amino acids is formed by the elimination of water between the COOH group of one amino acid and the NH_2 group of another, forming the covalent structure shown in figure 1.7. The peptide linkage secures the sequence of amino acids and gives the protein its primary structure, with the repeating unit $-CH-CO-NH-$ as its backbone. X-ray analysis has provided us with the representation of a peptide's primary structure shown in figure 1.8.

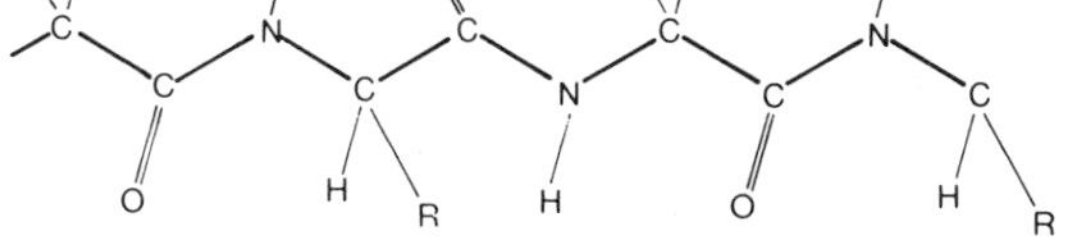

Figure 1.8 X-ray analysis of the peptide structure. (After Pauling et al., 1974.)

The peptide linkage leaves the protein molecule with two sources of electric charge: the COOH group at one end of the molecule and the NH_2 group at the other, and the charges on the R group (see figure 1.6). The net charge of a protein in solution depends on the pH of the solution. At its isoelectric point (IEP) the negative and positive charges on the protein chain keep each other in balance. If acid H^+ ions are added to such a solution, the negative groups will take them up, leaving the protein molecule with an overall positive charge. If alkali is added to a protein at IEP, the reverse takes place and the protein will be left with an overall negative charge. The size of the resulting charge will in both cases depend on the amount of ions added. The behaviour of a protein in acid and alkali solutions is as shown in a typical titration curve such as figure 1.9.

Other links which can add to the primary structure are brought about by the formation of covalent disulphide bridges. These bridges are less frequent than hydrogen bonds but much stronger. They are situated between peptide chains (interchain linking, see figure 1.11).

1.2.1.2 Secondary Structure

The linked polypeptides of proteins are usually coiled into a spiral known as a helix. The spiral is held together by hydrogen bonds (see section 3.7.2) between the NH group of one amino acid and the

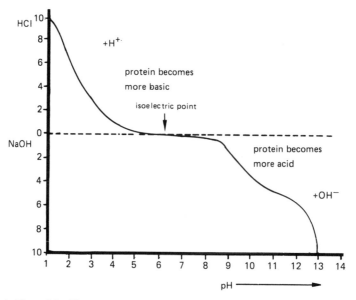

Figure 1.9   **The behaviour of a protein in acid and basic solution: a typical titration curve.**

oxygen atom of the CO group of another in the next turn of the helix. Although each bond is only weak, there are so many of them that together they form a bond strong enough to maintain the helix structure (see figure 1.10). In the case of the α-helix structure,

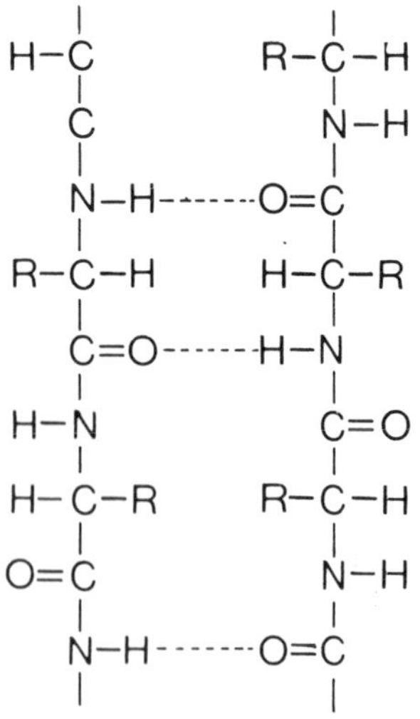

H–C   R–C–H
 | |
 C   N–H
 | |
N–H------O=C
 | |
R–C–H   H–C–R
 | |
C=O------H–N
 | |
H–N   C=O
 | |
H–C–R   R–C–H
 | |
O=C   N–H
 | |
N–H------O=C
 | |

Figure 1.10   **Interchain hydrogen bonding.**

hydrogen atoms are shared between carboxyl and amino groups of the two halves of the spiral loops facing each other. Another structure is provided by hydrogen atom sharing between the same groups of adjacent polypeptide chains. This gives the protein a pleated-sheet structure comparable to a Venetian blind (see figure 1.12).

The uncoiling of proteins is called denaturation and can be brought about in various ways, such as agitation, high-energy radiation, heating and freezing, solvents, oxidation, reduction and pH changes. Of these the first two do not play a part in cytological techniques, but the others are used to bring about changes in cell proteins (by fixation) in order to enable them to react with chemical compounds such as dyes.

The effect of pH, oxidation, reduction and solvents on the secondary structure of proteins is shown in figure 1.13. It is evident that denaturation will always lead to exposure of atoms or atom groups which before were concealed by the secondary structure. If these atoms or atom groups are charged, they will add to the reactivity of the protein and thus enhance staining.

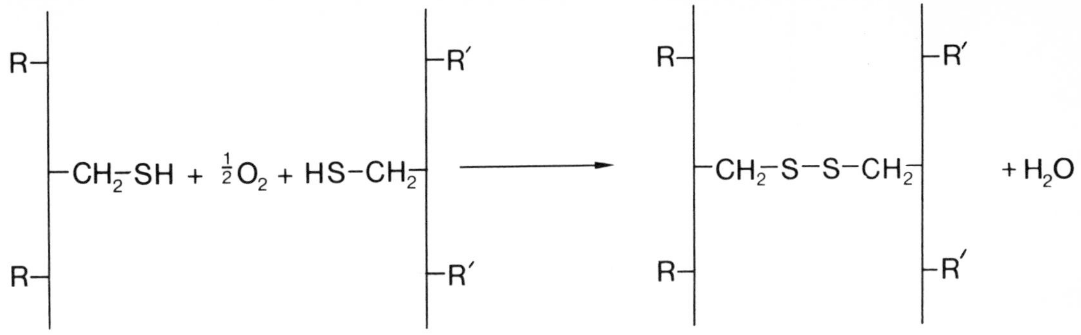

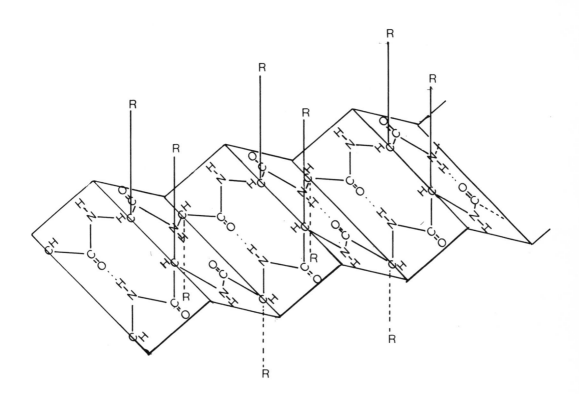

Figure 1.11   Interchain disulphide linking.

Figure 1.12   The pleated-sheet structure of proteins, picturing the structure of β-keratin as an example. For reasons of clarity the R groups which appear from underneath the pleated sheet have been omitted (after Zanker, 1981).

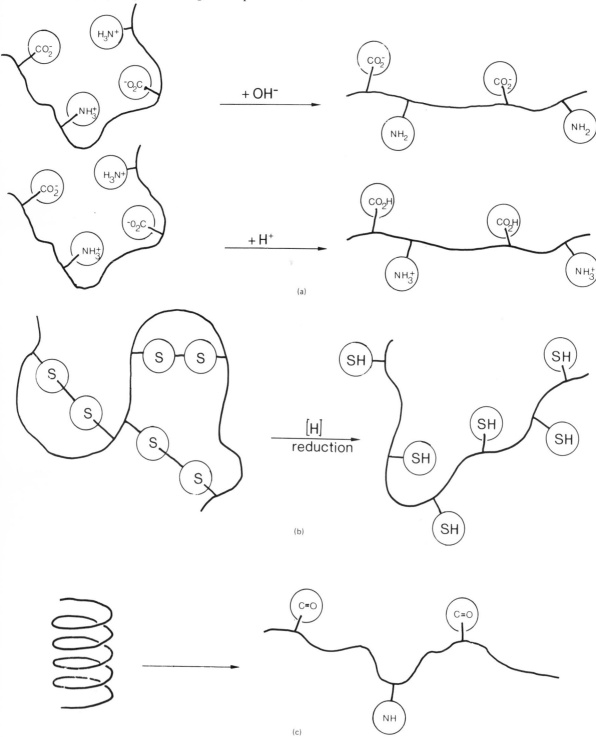

Figure 1.13   Denaturation of a protein by (a) destruction of the electrostatic forces by changing the pH of the solution; (b) destruction of the disulphide bond by reduction (after Ouelette, 1975); (c) by destruction of the hydrogen bond by the solvent.

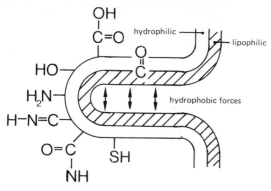

Figure 1.14 **The peptide chain, showing its hydrophobic and hydrophilic side chain (after Horobin, 1982).**

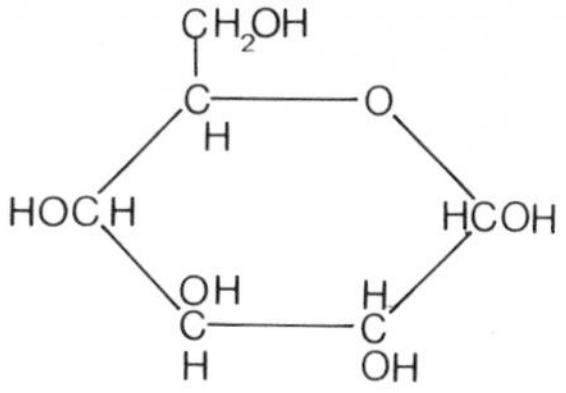

Figure 1.15 **Structure of an α-glucose molecule.**

1.2.2 Saccharides

1.2.1.3 Tertiary Structure

Native proteins can assume yet another form, the tertiary structure, in which the helix of the secondary structure is crumpled up into a more compact state instead of being straight. This is the case in globular proteins which form the majority of proteins. This 'crumpling' effect is caused by hydrophobic bonding (see figure 1.14, which shows part of a protein molecule in its tertiary structure). Groups which act as hydrophilic agents are the carboxyl, hydroxyl and amino groups at the side chains, the carbonyl groups in the backbone, the amino groups of several amino acids, the amino groups of asparagine, and the sulphydryl group of cysteine. Long fibrous protein molecules lacking this tertiary structure are found in the keratin of hair, collagen of tendon and bone, and myosin of muscle.

As with the secondary structure, the undoing of the tertiary structure also plays a part in altering the reactive properties of cell proteins and thus enhances staining.

1.2.2.1 Introduction

Saccharides are carbohydrates consisting of smaller or larger molecules with the general formula $C_x(H_2O)_y$. The single units, the monosaccharides, can be linked together with the elimination of water forming di-, tri-, and polysaccharides. The monosaccharides can take on several structural forms and can be made up of different numbers of carbon atoms, ranging from three to seven. Those single units that form the base for the formation of polysaccharides always have six carbon atoms. The saccharides are soluble in water and form sources of energy for all organisms.

1.2.2.2 Glucose

Glucose (figure 1.15), a monosaccharide, is the basic unit of the polysaccharides, with six carbon atoms of which five have hydroxyl groups and one an aldehyde residue (CHO). Saccharides with this structure are called aldoses. Fructose, which is also a monosaccharide, differs from glucose in that it does not have an aldehyde group but a keto group (CO). Hence the name ketoses, to which deoxyribose and ribose belong. Glucose forms the primary energy source for the cell.

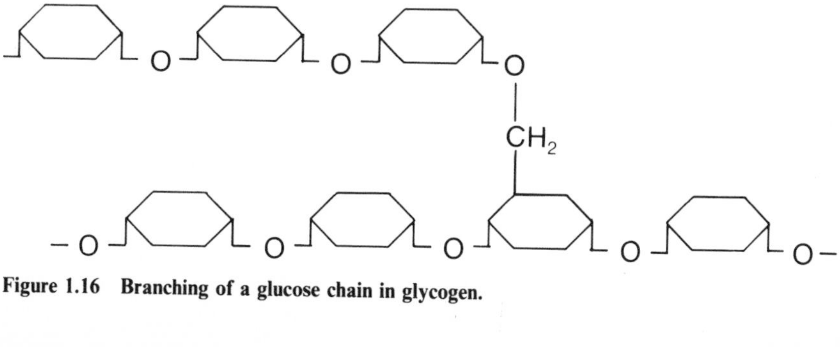

Figure 1.16 **Branching of a glucose chain in glycogen.**

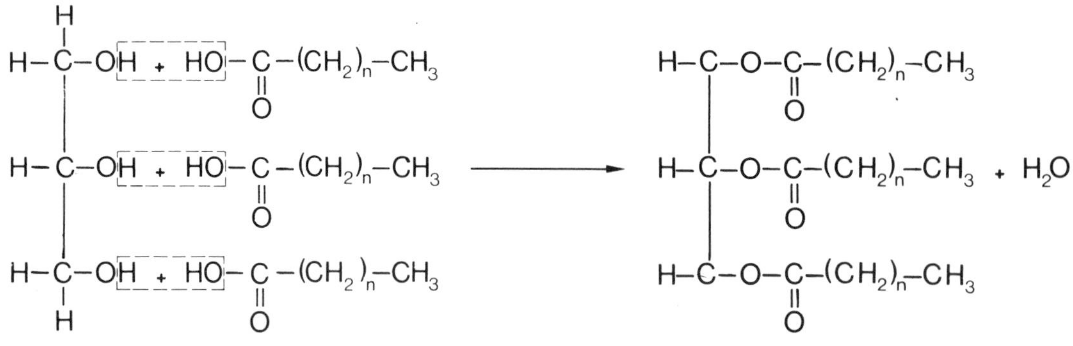

Figure 1.17   Formation of a triglyceride.

1.2.2.3 Glycogen

Glycogen is a polysaccharide which can be found in many cells as a energy reserve, especially in liver and muscle cells. It is built up from many glucose molecules not in a single chain (like, for instance, cellulose) but in a branched structure (see figure 1.16). The aldehydes on the glucose molecule after oxidation give glycogen the ability to react with Schiff's reagent (see section 3.5.2). Glycogen can also be dyed by Best's Carmine (see section 3.6).

1.2.2.4 Glycoproteins

Simple or more complicated saccharides can form complexes with proteins to form glycoproteins. Cellular glycoproteins can be found in the cell membrane as mucin with the carbohydrate end pointing to the outside of the cell. They play an important role in membrane interaction and recognition. Here again, Schiff's reagent can be used to demonstrate the aldehydes and thus the mucin. The polysaccharides in mucin can also be dyed by Alcian Blue (see appendix 2). Secretory glycoproteins are produced by many different kinds of cells, e.g. those of the liver, thyroid and pancreas (deoxyribonuclease and ribonuclease), plasma cells and others.

1.2.3 Lipids

1.2.3.1 Introduction

The solubility of lipids in organic solvents but not in water is caused by their long aliphatic hydrocarbon chains or benzene rings. Most lipids are therefore non-polar and hydrophobic. They are esters, mostly of glycerol and fatty acids, and are usually divided into two groups: the triglycerides and compound lipids.

1.2.3.2 Triglycerides

Triglycerides or neutral fats are accumulated in the cytoplasm of fat cells as an energy reserve. In addition, they can be found in the cytoplasm of other cells. They are esters of one glycerol molecule and three fatty acids, as shown in figure 1.17.

Oxidation of triglycerides releases a good deal of energy. Hydrolysis separates glycerol from the fatty acids. If there is a double bond in the fatty acid chain between two carbon atoms, the fat is described as unsaturated.

Because of their hydrophobic character, fats cannot be dyed. How they can be visualised in other ways is shown in section 3.5.3. Dyes which are commonly used to stain fats are Sudan Black, Oil Red and Luxol Fast Blue (see appendix 2).

1.2.3.3 Compound Lipids

Compound lipids are part of the structure of membranes in the cell. Unlike the triglycerides they are partly polar. They have a hydrophobic 'head' and two hydrophobic 'tails' (see figure 1.18a). For the phospholipids, for instance, this comes about as the result of esterisation of the third hydroxyl group of glycerol with phosphoric acid. The phosphate which is thus formed is bound to a second alcohol which can be choline, ethanolamine, inositol or

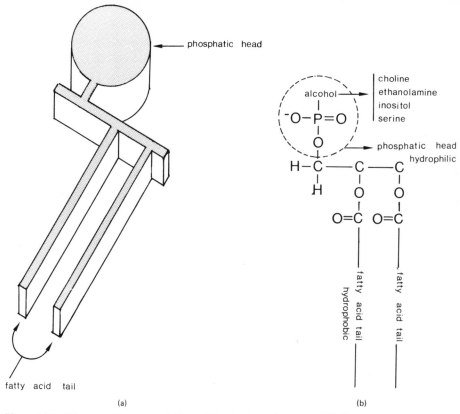

choline
ethanolamine
inositol
serine

alcohol

$^{-}O-P=O$

phosphatic head
hydrophilic

$H-C \longrightarrow C \longrightarrow C$

phosphatic head

fatty acid tail
hydrophobic

fatty acid tail

(a)   (b)

Figure 1.18   Diagrammatic representations of the structure of compound lipids.

serine. This phosphate head is polar and thus hydrophilic, and the fatty acid chains remain hydrophobic (see figure 1.18b). In biological membranes these hydrophilic heads form the layer which faces water on the outside. The hydrophobic tails point to the interior of the membrane structure. Lecithin, cephalin and plasmalogen belong to this group of compound lipids. Plasmalogen contains long-chain aliphatic aldehydes which can give a direct reaction with Schiff's reagent. In other compound lipids, glycerol is replaced by more complicated alcohols, but they all have in common the fact that they have a polar head and a hydrophobic tail.

Table 1.2   Reactivity of the chemical constituents of the cytoplasm

Constituent	Reactive group
Proteins	Depends on which atoms or atom groups are exposed through fixation
Saccharides	1,2-glycerol group after oxidation (see section 3.5.2.3)
Lipids	None, for colouring see section 3.5.3
RNA	Phosphate

2

Comparing Cells in Histology and Cytology

2.1 Introduction

Although a knowledge of histopathology is of
paramount importance in diagnostic cytology
(Koss, 1979), it is far too simple to state that the
appearance of cells in a cytological specimen is
identical to that of the corresponding cells in
histological material. There are several reasons for
the presence of differences in cell images in histology and cytology, and of differences in cell
harvest in the two methods.

(1) The cell may change when it is removed from
    the tissue.
(2) As a result of the process of reduction from
    three dimensions to two, the shape of the cell
    and its size are altered in a different way in the
    two methods. In histological material this re-
    duction is achieved by cutting the cells. In
    cytological specimens it is brought about by
    spreading the cells over the glass slide.
(3) When the cell exfoliates in the body fluids, and
    remains floating for some period of time, its
    morphology may change significantly.
(4) All the cell types present in the tissue are
    visualised in histology, and cell loss does not
    occur. In contrast, in cytology, preferential
    sampling of cell types takes place and cells will
    have different properties. Moreover, loss of
    cells during cytopreparation may occur. These
    mechanisms may lead to differences in cell
    harvest in the two methods.

The differences between the images of cells in
histology and cytology and the underlying causes
will be further elucidated in this chapter. In addition, the differences in staining histological and
cytological slides will be discussed. The techniques
for cytopreparation are described in chapter 10.

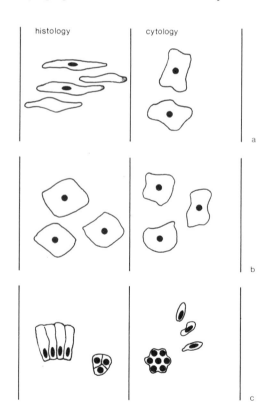

Figure 2.1 Cells in cytology and histology. (a) Superficial
squamous cells. (b) Intermediate, glycogen-con-
taining, squamous cells. (c) Cylindrical endocervical cells.

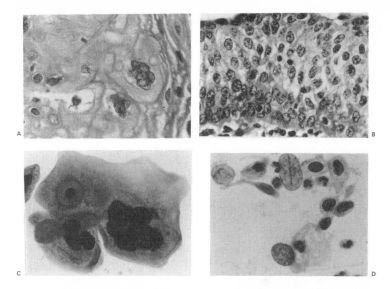

Figure 2.2   Dysplastic and atypical reserve cells in cytology and histology. (A) Dysplastic cells in histology; note dense cytoplasm. (B) Atypical reserve cells in histology; note cloudy, partly disrupted cytoplasm. (C) Dysplastic cells in cytology with intact cytoplasm. (D) Atypical reserve cells with only remnants of cytoplasm. Note the 'strand' in the nucleus due to deflation.

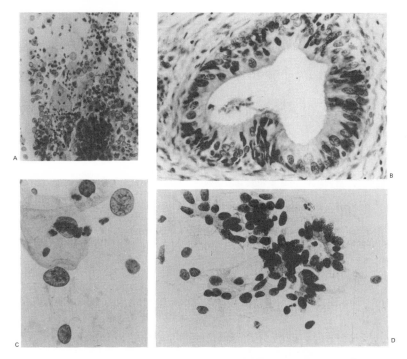

Figure 2.3   Atypical reserve cells and cells from well differentiated adenocarcinoma (in situ) of the endocervical epithelium. (A) Exfoliative pattern of atypical reserve cells. (B) Tissue section of adenocarcinoma in situ. (C) Atypical reserve cells; note deflated nuclei with 'chromatin strands'. (D) Cells from adenocarcinoma in situ; note intact cytoplasm (cytology).

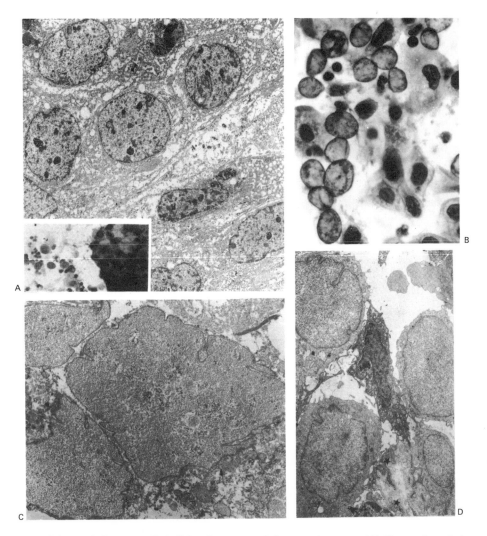

Figure 2.4 Atypical reserve cells in light microscopy and electron microscopy. (A) Cluster of atypical reserve cells taken from smear; note ill-defined although intact cytoplasm (EM) (inset: cytological smear from which cluster of atypical reserve cells (right) is taken for EM). (B) Atypical reserve cells in smear. (C) EM of atypical reserve cells taken from smear; note loss of cytoplasm and intact nuclear membrane. (D) EM of histological material; note resemblance to nuclei in C.

2.2 Cell Changes Occurring when Cells are Removed from Tissues

Through scraping (cervical cytology), brushing (lung cytology), or aspirating (fine needle aspiration cytology), cells are removed from the tissue. Depending on the physical and chemical properties of the cytoplasm, the latter might be damaged in each of these mechanical processes. For instance, the firm cytoplasm of well differentiated squamous cells withstands such forceful processes. These cells will keep their cytoplasm, although for cell removal the integrity of the desmosomes must be broken. On the other hand, cells loaded with fat, and cells with poorly differentiated cytoplasm, are easily damaged. In aspirating cells from fatty liver, many

Figure 2.5   EM of nuclei of atypical reserve cell and dysplastic cell taken from smear. (A) Nucleus of atypical reserve cell; note intact nuclear membrane. (B) Nucleus of dysplastic cell; note incomplete nuclear membrane.

cells lose their cytoplasm and thus the smear will contain a great number of bare nuclei. In histological sections, the cytoplasm of these cells is not so easily disrupted.

In contrast, the cytoplasm of poorly differentiated epithelial cells also shows damage often in histological sections. The cytoplasm is then seen either as cloudy material, or as spaces containing threads (see figure 2.1). The amount of cytoplasm can then be calculated from the distance between the nuclei lying in the open spaces. This phenomenon is seen in sections of both benign and malignant poorly differentiated epithelium. When these poorly differentiated cells are removed from the tissue, their cytoplasm is partly or completely lost (figures 2.2 to 2.5).

We have shown in earlier work (Ruiter *et al.*, 1979) that immature benign cells from the cervical epithelium (reserve cells), when found in cervical smears, have light-microscopically bare nuclei. However, EM study of the same cells visualised that this loss of cytoplasm was not due to the fact that these cells were degenerated, because they had an intact nuclear membrane (figure 2.1). It was suggested that their fragile cytoplasm was disrupted by the taking and smearing of the cell material.

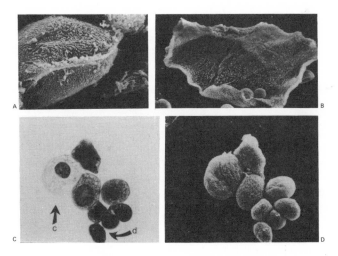

Figure 2.6   SEM of benign squamous cells and malignant cells in smear. (A) Thick intermediate cell. (B) Flat superficial cell. (C) Light microscopy of malignant cells (c) and dysplastic cell (d). (D) SEM of malignant cells and dysplastic cell (see C).

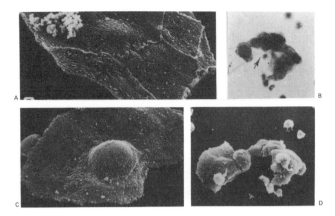

Figure 2.7   **SEM of abnormal cervical cells. (A) SEM of squamous cell with slightly enlarged nucleus. (B) Light microscopy of abnormal immature cells. (C) SEM of squamous cell with large nucleus. (D) SEM of cells shown in B.**

The same is valid for poorly differentiated malignant cells (Spaander *et al.*, 1982). Remnants of the disrupted cytoplasm still containing mitochondria can be found around the nucleus. In addition the shape of the nuclei of these poorly differentiated cells is easily altered by the smearing process (see section 2.3.4). The nuclear shape of differentiated cells, protected by their intact cytoplasm, is seldom altered by the smearing process. Moreover, the nucleus of poorly differentiated cells easily collapses, causing changes in the chromatin pattern (see section 2.3.2 and figures 2.2 and 2.3, deflation).

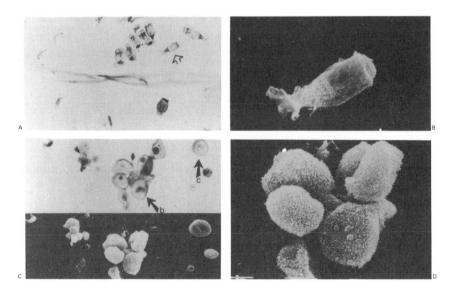

Figure 2.8   **SEM of cylindrical endocervical cells and metaplastic cells. (A) Cylindrical mucus-producing endocervical cells. (B) SEM of cell shown (with arrow) in A. (C) Light microscopy and SEM of the same metaplastic cells. b and c are metaplastic cells. The other cells are mucus-producing endocervical cells and leucocytes. (D) Enlargement of C.**

2.3 Histological Cutting versus Cytological Spreading

In light-microscopical study of cells it is necessary to obtain two-dimensional objects. In histological techniques the reduction from three to two dimensions of the cell and its nucleus is achieved by cutting the embedded cell in thin slices. In cytology this reduction is accomplished by means of the process of spreading the cell over the glass slide. Here the cell is not cut but kept as a whole.

In histology, the angle of cutting the cell will be decisive for the resulting cell image. For instance, if the cell is spherical, all cell images will be circular. However, if one of the three cell dimensions differs significantly from the other two, such as is the case in flat squamous cells, different cell images may occur. The majority of flat squamous cells will be thin and elongated in histological sections. In contrast, squamous cells in cytology specimens are large and flat, because they spread over the glass slide on their large, flat side (figure 2.1). This is visualised in the SEM photographs (figure 2.6), in which the squamous cell lies flat with the nucleus protruding from the plane of the cytoplasm, (figure 2.7). Thus, different parts of the cells in cytology specimens might vary in thickness. It is clear that the cellular images of flat squamous cells in histology and cytology differ greatly. When the squamous cells are not flat, but thick due to a high glycogen content, such as is the case in pregnancy, or are thick because of immaturity (metaplastic cells, see figure 2.8), the cellular images in histology and cytology differ less (see figure 2.1).

Cylindrical endocervical cells mostly lie on their long side on the slide in cytology specimens (figures 2.8 and 2.9), and, in contrast to the nuclei of the flat squamous cells, their nuclei do *not* protrude. Only when the cylindrical cells are arranged in cell groupings do they lie with their short side on the glass surface. The differences in cellular image of the cylindrical cells in histology and cytology are less dramatic compared with those of the flat squamous cells.

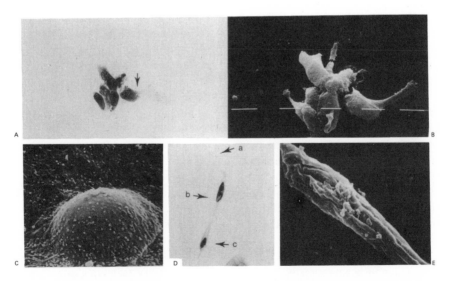

Figure 2.9 Nuclear protrusion. (A,B) Light microscopy of abnormal endocervical cells; no nuclear protrusion; it is visible in the SEM (B) that these cells have an abnormal shape, which in light microscopy is less pronounced (see arrow in A). (C) Pronounced nuclear protrusion of a dysplastic cell. (D,E) Light microscopy of squamous spindle cells from carcinomas (a, b and c in D); no nuclear protrusion.

2.3.1 Spreading of Cells over the Glass Slide

When a cell is placed on a microscope slide, the nucleus and the cytoplasm spread over the glass. One can compare this with the spreading of an egg in a frying pan. In the spreading process, both the nucleus and the cytoplasm will become flatter. This flattening process is essential for light-microscopic study of cells in cytological specimens. In the process, the nuclear and cytoplasmic areas will be enlarged, just as is the case with the egg in the frying pan. The spreading process is found not only in single cells, but also in cell groupings.

The alteration of the cells in the spreading process is not the same for all cell types, and is dependent on the following cell properties.

(1) Nuclear size and rigidity. Large nuclei are modified much more than small nuclei, and flexible vesicular nuclei spread much more than rigid pyknotic nuclei.
(2) Cytoplasmic size and rigidity. Large cells are modified much more than small cells, and flexible, poorly differentiated cytoplasm spreads much more than rigid well differentiated cytoplasm. The squamous cell is a type of cell that hardly spreads. In addition, different parts of the cell might differ in rigidity. An example of this can be seen in mesothelial cells. The laminar outer zone of the cytoplasm of these cells spreads much more over the glass slide, and thus is also much thinner, than the rigid inner zone.
(3) Fixation and pre-fixation method. Cytological size modifications are greatest when the cells are in a monolayer and are in addition fixed by air drying and well separated from each other. Cells in cell groupings and superimposed cells are blocked in their spreading, as are cells that come into contact with each other in the cell spreading process (see also section 2.5).
(4) Cell environment. The conditions that surround the cell influence its spreading. If the cell is dried in a mucoid or protein-rich environment, the spreading is sub-maximal. It is possible that it is not only the mere physical presence of the mucoid and protein material that inter-

feres with the spreading of the cell, but in addition the denaturation of the proteins in it during fixation.

The effects of the mechanisms discussed above are most prominent when the cells are fixed by air drying, as in the Romanowsky–Giemsa methods. However, for the Papanicolaou staining methods, the cells are fixed in alcohol solutions (see chapter 4) and these effects are less. As is discussed in detail in section 4.2.2, alcohol stiffens the nucleus and cytoplasm by denaturation of its proteins, and, in addition, causes nuclear and cytoplasmic shrinkage. Alcohol fixation can be performed by putting the fresh slide in an alcohol bath or spraying it with a spray fixative containing alcohol. These are described as wet-fix methods. So, in the case of a freshly made cell smear subsequently put into alcohol, there are two competing processes: the process of cellular spreading with additional increase in nuclear and cytoplasmic area, and the process of nuclear and cytoplasmic hardening and shrinkage. The hardening blocks the change from a three-dimensional object to a flat one. The final product is the result of the two opposing mechanisms. The influence of the hardening and shrinkage effect depends on the grade and concentration of the alcohol used and its combination with other components in the spray fixative. The higher the percentage of alcohol, the more severe is the shrinkage. Spray fixative containing polyethylene glycol gives a slightly less severe increase in nuclear and cytoplasmic size. Figure 2.10 gives the results for nuclear and cytoplasmic sizes of benign urothelial cells prepared in various ways. The variations between the results in the different cytopreparatory techniques are not identical for other cell types. It is a regrettable fact that there does not exist one general formula to calculate the cellular and nuclear sizes resulting from the various wet-fix and air-dry techniques.

The almost completely flattened cells produced in the air-dry technique adhere well to the glass surface. To a certain extent the spherical cells flatten in the wet-fix methods; completely spherical cells cannot adhere to the glass surface and roll like marbles from the slide.

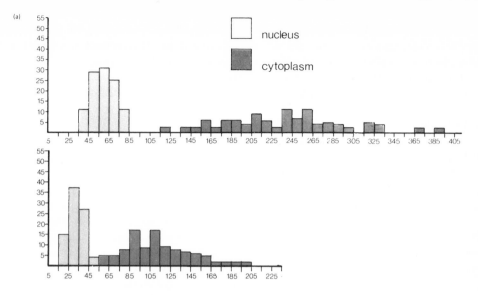

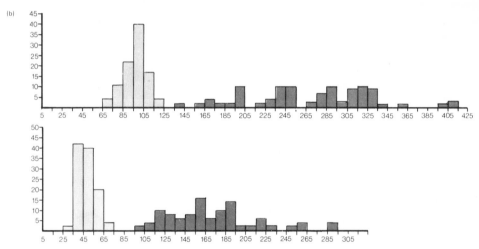

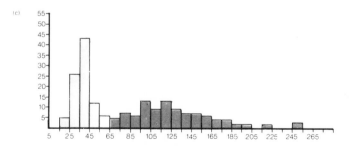

Figure 2.10   Nuclear and cytoplasmic sizes of benign urothelial cells in various cytopreparatory techniques. *x*-axis, size in nm²; *y*-axis, percentage of cells. (a) Top: no fixation, no staining; bottom: 96% ethyl alcohol, Papanicolaou. (b) Top: air-dried May–Grünwald--Giemsa (MGG); bottom: 50% ethyl alcohol, Papanicolaou. (c) Leiden spray fixative, Papanicolaou.

A separate story is the behaviour of cells fixed in suspension prior to slide preparation. This happens in the Cytospin method (section 10.4.2) and in the Saccomano technique (section 10.2.3.2): polyethylene glycol–alcohol solutions are then added to the fresh cells. In the Cytospin method, the prefixed, slightly rigid cells adhere to the glass slide due to the fact that they are flattened by the mechanical forces during centrifugation. In the Saccomano technique the cells are glued to the glass slide by the polyethylene glycol, which is also an additional mechanism in the Cytospin method. In the former, the three-dimensional cellular and nuclear changes are minimal, and so spherical structures remain almost spherical.

The properties of spherical cells differ in the air-dry and wet-fix methods (see figures 2.11 and 2.12). The shape of the spherical cells prepared fresh and wet-fixed on the slides can be compared to poached eggs on a piece of toast. Fresh cells air dried on the slide can be compared to fried eggs. Cells prefixed in suspension with alcohol are like hard-boiled eggs pressed with force on a piece of toast. From figure 2.11 it is clear that in the wet-fix techniques the nucleus can be quite thick, whereas in the air-dry methods the nucleus is invariably flat. This explains why, in light microscopy, on different levels of focusing using high power ($\times 475$ or more), different chromatin patterns may exist. This phenomenon is absent in the flat air-dried nuclei, which are only 'in focus' on one focusing plane. In the thick, wet-fixed nuclei, by changing the focusing plane the nucleoli can be visualised on different locations. In contrast in the flat air-dried nuclei, all nucleoli are simultaneously visible in one focusing plane.

The changes described above also occur in cell groupings. This implies that in the wet-fix techniques the cell groupings are much thicker than in the air-dry techniques. This can cause problems in the light-microscopical study of cell groupings. Also the staining results in cell groupings are

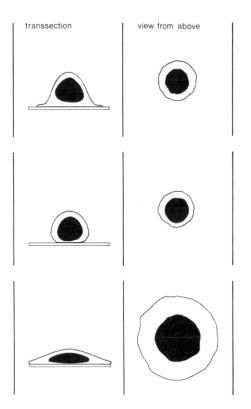

transsection   view from above

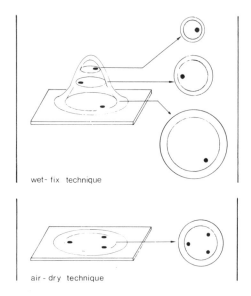

wet-fix technique

air-dry technique

Figure 2.11   Spherical cells in various cytopreparatory techniques. (a) Cell prepared fresh and wet-fixed on the slide. (b) Pre-fixed cell in suspension, deposited on slide. (c) Fresh cell air dried on slide.

Figure 2.12   Nucleolar localisation in various cytopreparatory techniques: wet-fix technique (three focusing planes) and air-dry technique (one focusing plane).

influenced by their thickness (see Atlas, plates 6.4, 19.3 and 24.4).

Furthermore, when the cell changes from a three-dimensional object to a two-dimensional one, the constituents of the cytoplasm may change in location and distribution. For instance, lipid vacuoles of mesothelial cells in histological sections are equally divided over the cytoplasm of the spherical cells. In the flat, air-dried cells, redistribution of the vacuoles has resulted in their perinuclear location (figure 2.13). This is caused by the fact that the vacuoles are positioned in the thickest, middle part of the cell. In the wet-fix techniques, the lipid vacuoles collapse because the lipids are extracted from the cytoplasm by the alcohol and thus the vacuoles are virtually invisible (see Atlas, plate 16.3).

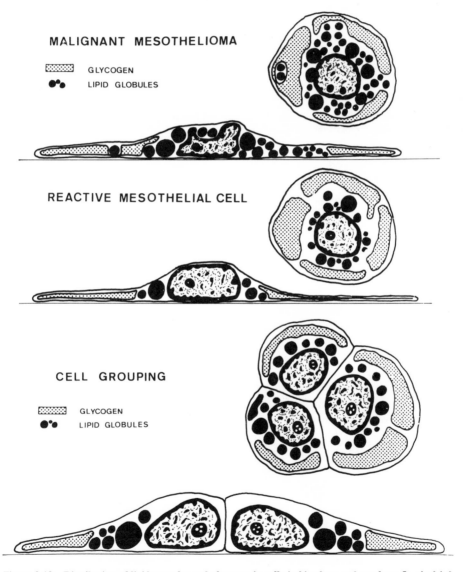

Figure 2.13   Distribution of lipid vacuoles and glycogen in cells in histology and cytology. In air-dried cytology smears (top), the lipid vacuoles are in the thickest part of the cell (perinuclear). In cell groupings (bottom) the distribution is slightly different.

2.3.2 Effect on the Nuclear Image when the Nucleus Collapses

When the nucleus is thick and flexible, it may collapse in the wet-fix techniques. As a consequence it lies on the microscope slide like a deflated balloon. This is particularly the case in poorly differentiated cells with large vesicular nuclei. The 'deflated' nucleus will lie in folds, and the thus-folded nuclear membrane with its chromatin will cause a nuclear image with strands (see figures 2.2 and 2.3). In contrast, the rigid nuclei of squamous cell carcinoma will not deflate (see figure 2.4).

2.3.3 Changes in Nuclear Shape

2.3.3.1 Changes in the Cytospin Methods

It is a well known fact that, when the nucleus has a gyriform shape, the nuclear profile in the histological section is dependent on the level of sectioning and the thickness of the section. This is visualised in figure 2.14; lymphocytes (a certain sub-population thereof) in thick sections (6 nm) show only small, shallow indentations of the nucleus. In the thin sections (1 nm), it becomes visible that these nuclei actually have very deep indentations, resulting in an almost gyriform shape. The nuclear shape in cytology specimens of such cells is close to that in thick histological sections. However, especially for cells with one deep indentation, such as cleaved cells in malignant lymphoma, the nuclear cleavage can only be visualised when the cells are deposited on the glass slide with force. This is the case, for example, in the Cytospin methods. In addition nuclear cleavage is more prominent in the air-dry methods (see figure 2.15). A golden rule is: if one wants to see nuclear indentations, one should choose the technique that flattens the nucleus most.

2.3.4 Changes in the Smear Methods

When a smear is made of the cells (see chapter 10), the nuclear shape can be altered by the mechanical force of, for instance, the wooden spatula. This is especially the case when the nucleus is not protected

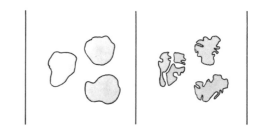

Figure 2.14 Nuclei in 6 μm sections (left) and ultrathin sections (right). In the former, the nuclei show only shallow indentations. In ultrathin sections the nuclear shape is gyriform.

by well differentiated cytoplasm; for example, in cells with ill-defined, cloudy cytoplasm, such as is seen in histiocytes, or in poorly differentiated cells.

2.4 Changes in Floating Cells

When the cells are exfoliated in body fluids and remain floating for a long period of time before being harvested, they may develop dramatic morphological changes. As an example we will discuss the changes occurring in mesothelioma cells. In our previous studies (Boon *et al.*, 1982*a*, 1984*a*), we found that the exfoliated malignant mesothelioma cells mature while floating. The same signs of maturation are seen in cultured malignant mesothelioma cells. This type of maturation, however, is much less pronounced in malignant mesothelioma cells in tissue (Boon *et al.*, 1981). This form of maturation encompasses predominantly lipid

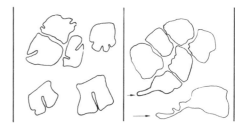

Figure 2.15 Cleaved nuclei in the Cytospin method and the smear method. Left: in the Cytospin method the cleavage is clearly visualised. Right: in the smear method this is not the case; in addition the nuclei show smearing damage (arrows).

production. These lipids are located in vacuoles, which in the first stages of maturation are found to be perinuclear (see figure 2.16). In later stages, the lipid vacuoles become more numerous, especially in malignant mesothelial cells (Kwee *et al.*, 1982). Especially when the primary site of the mesothelioma is peritoneal, lipid production in the floating cells can be extreme, giving the malignant cells a 'benign' appearance due to their very low nuclear to cytoplasmic ratio (Boon *et al.*, 1981).

The lipid-laden malignant mesothelial cells lie mostly single or in twos, whereas the less mature malignant cells, with little lipid, are found in dense clusters. These clusters are heavier than the lipid-laden isolated cells. Therefore, when the cells are aspirated in the upper parts of the exudate, the single cells predominate, and when the aspiration is performed in the lower part of the exudate, the slides will only show immature cell clusters (see figures 2.16 and 2.17). The cytological pattern of the two aspirates will then be completely different. The same thing happens when exudate is left overnight—the upper part will contain the lipid-laden mesothelioma cells, and the lower part the heavy dense clusters without lipid.

In general, cells in cysts will also change, becom-

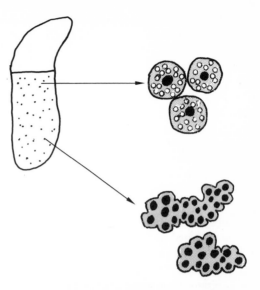

Figure 2.17 **Preferential cell sampling in pleural fluid. Above: aspirate taken from the upper part of the fluid; mature malignant mesothelial cells lie isolated, are lipid-rich and large. Below: aspirate taken from lower part of the fluid; malignant mesothelial cells in dense clusters (these cells contain small amounts of lipids and are small).**

ing very vacuolated. This is, for instance, the case in cells from Nabothian cysts of the cervix (Boon *et al.*, 1981) and from breast cysts (Zajicek, 1974) and prostate (Zajicek, 1979).

The changes in floating cells described above can cause dramatic differences in cell morphology of the same type of cells in histological sections and cytological material.

2.5 Cell Harvest in Histology and Cytology

2.5.1 Preferential Sampling

Preferential sampling in cytology can occur as a result of the physical characteristics of cells floating in fluids, as is described in section 2.4. In addition, when cells are removed from the tissue, the rate of sampling depends on their anchorage in the mother tissue. When the cells are sampled from tissue of a mixed cell population, those cells will predominate in the cytological slides which are most easily removed from the tissue. This phenomenon also results in preferential sampling.

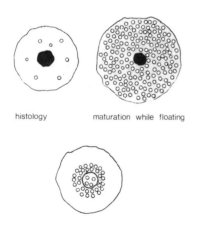

Figure 2.16 **Lipid production of malignant mesothelial cells in histology, in which little lipid production is observed, and in cytology, when peri- and supranuclear lipid vacuoles can be seen. The cells mature while floating, with concomitant increase of lipid vacuoles.**

2.5.1.1 Aspiration Cytology

The most impressive example of preferential sampling can be found in aspiration cytology. Aspiration cytology is known by many different names: aspiration biopsy cytology (ABC); thin needle aspiration cytology (TNA); thin needle aspiration biopsy; needle aspiration biopsy; and fine needle aspiration biopsy cytology (FNA). This reflects the fact that confusion exists about the mechanism behind aspirating cells with a thin needle. Thompson (1982) argues that the cells are mainly dislodged by the negative pressure produced in the needle, and not by the 'cutting' action of it. Whether a cell will be aspirated or not is then completely dependent on its anchorage in the mother tissue. Thompson's theories are strengthened by the observation that aspirates from tissue containing a mixture of epithelial and mesenchymal cells, such as the prostate, contain almost exclusively epithelial cells (which are easily dislodged), and very few mesenchymal cells (which are strongly bound). If a histological section is made from a biopsy, taken with a thick ('truecut') needle, this is not the case. We therefore prefer, like Trott (1983), not to use the word 'aspiration biopsy' if cytological aspiration with a thin needle is meant.

In aspiration cytology this preferential sampling can aid in the diagnosis of malignancy: carcinoma cells are less cohesive compared with benign epithelial cells, resulting in more cellular aspirates. However, in some fields of cytodiagnosis the opposite may be true: the epithelioid cells, on which the cytological diagnosis of Morbus–Besnier–Boeck is based, are more difficult to dislodge than the surrounding lymphoid cells in the aspirated lymph node. Therefore the diagnostic epithelioid cells are scarce in cytological material in comparison with histological material.

2.5.1.2 Other Cytology

When the cells are removed by scraping or brushing, or when they exfoliate spontaneously, the cohesiveness of the cells is a key factor determining their presence in the cytology specimen. Here the decreased cohesiveness of carcinoma cells aids us in cytodiagnosis. However, when quantitative studies on cell populations are performed, one should be aware of the fact that frequently the proportions of various cell types in histological and corresponding cytological specimens differ significantly.

2.5.2 Cell Loss during Slide Preparation and Staining

Cell loss during slide preparation and staining plays an important role in cytology, but it hardly occurs in histology. In the first place it is of importance that the cells adhere to the glass slide, and in the second place that they stay on it. Cell adherence to the glass slide is dependent on several factors:

(1) Unfixed cells adhere better than fixed ones. Fixed cells are more rigid, and may roll like marbles from the smooth surface of the glass slide.
(2) Air drying promotes cell adherence (see section 2.2).
(3) Wet fixation by putting the freshly made smear in 95% ethyl alcohol diminishes cell adherence.
(4) Polyethylene glycol (in spray fixatives) glues the cells to the glass surface: when it is removed afterwards, the cells stay on the slide.
(5) Mucus and proteins in the cell specimen promote cell adherence.
(6) Smooth glass surfaces of glass slides diminish cell adherence; thus cells stick better on frosted slides.
(7) Cells stick well on very cold glass slides taken from the freezer (see chapter 10).
(8) Slides coated with adhesives retain cells better (see section 10.8).
(9) Mechanical forces promote cell adherence. This is for instance the case when the sample is centrifuged using a special bucket (the Leif bucket) in which the cells are directly deposited on the slide. When the supernatant is decanted after centrifugation, the cells stay on the slide (see section 2.5.3).
(10) In the Cytospin method, cell loss can occur when the cell sample is absorbed by the filter paper prior to centrifugation (Beyer Boon *et al.*, 1979a). This cell loss can be limited if, by

Table 2.1 Cell loss during slide preparation and staining. Cells from urothelial carcinomas (from Beyer-Boon *et al.*, 1978, *Acta Cytologica*, **22**, 589–94)

Case numbers*	1	2	3	4	5	6	7	8	9	10
Number of cells per ml	49 200 100%	51 500 100%	49 900 100%	98 800 100%	95 000 100%	82 200 100%	298 100%	1 245 100%	1 240 100%	1 309 100%
A Wet fixation albuminised slides, Pap staining†	1 887 4%	3 197 6%	6 353 13%	—	—	—	40 13%	65 5%	71 6%	46 4%
B Wet fixation no albumin, Pap staining	1 980 4%	4 465 9%	5 082 10%	—	—	—	53 18%	72 6%	130 11%	107 8%
C Wet fixation alc. 96% Pap staining	1 501 3%	1 075 2%	927 2%	18 772 19%	14 250 15%	18 084 22%	131 44%	85 7%	237 19%	211 16%
D Spray fixation Pap staining	9 938 20%	21 458 42%	19 312 39%	53 328 56%	43 700 46%	13 152 16%	179 60%	457 37%	572 46%	241 18%
E Air drying Pap staining	21 195 43%	26 498 51%	35 591 71%	61 256 62%	58 900 62%	52 608 64%	187 63%	1 025 82%	1 298 105%	1 090 83%
F Air drying albumin MGG staining**	28 481 58%	40 555 79%	43 339 87%	85 956 88%	80 732 85%	53 430 65%	296 99%	925 74%	1 059 85%	1 030 79%
G Air drying MGG staining	47 278 96%	40 311 78%	40 482 81%	75 088 76%	78 850 83%	62 472 76%	269 90%	962 77%	1 090 88%	1 020 78%
H Millipore filter Pap staining	40 195 82%	46 842 91%	42 437 85%	88 920 90%	80 750 85%	53 395 65%	274 92%	1 182 95%	916 74%	925 71%

*1–6 = Grade 2 and 3 carcinomas; 7 = Grade 1 tumour; 8–10 = Negatives.
†Pap = Papanicolaou.
**MGG = May-Grünwald–Giemsa.

the technique of pipetting the aliquot in the cuvette, an air bubble is created separating the sample from the slide (Boon *et al.*, 1983). In the new Cytospin II, this air bubble is created automatically due to the angle in which the cuvette is placed.

From the above list it is clear that the cell harvest is dependent on the state in which the cells are prepared (fixed, unfixed), the method of fixing the slide, the type of specimen (body fluids containing little protein or mucus have dramatic cell losses), and the type of slide used (coated, prepared). The large differences resulting from the various cytopreparatory techniques for urine are shown in Table 2.1.

It is evident that cells that are poorly attached to the glass slide can also easily be detached in the staining or alcohol bath, and thus cause 'floaters'. Floaters are cells or cell groupings that detach from the glass slide, float in the (staining) bath, and consequently attach to another slide. This is especially dangerous when malignant cell groupings become attached to a slide from a patient without cancer. This might happen when urinary specimens or slides from pleural fluid containing malignant cells are processed, and may then lead to erroneous diagnoses. Floaters lie on another focusing plane than the rest of the cells in the slide, and can thus be recognised.

2.5.3 Concentrating Cells from Cell Suspensions

Cells in suspension (coelomic fluids, CSF, urine) must be concentrated for slide preparation. Several techniques are used for this purpose:

(1) Preparing a cell pellet in the centrifuge, decanting the supernatant and using the cell pellet to make smear preparations.
(2) Concentrating the cells with a centrifuge directly *on* the slide. This system is used in cytocentrifuges, such as Cytospin I and II, or by using special buckets in the conventional centrifuge (Leif *et al.*, 1977).
(3) Concentrating the cells by gradient techniques: the cells are caught in different solutions with

various densities; smear preparations can be made from each layer containing different cell populations. In chapter 10 we describe the percoll gradient method.
(4) Concentrating cells by the sedimentation methods, which are very gentle.

All four methods are described in section 10.4.

It is beyond the scope of this book to go deeply into the advantages and disadvantages of the above-mentioned cell concentration techniques, and the reader is therefore advised to consult the relevant literature. It is clear that for each type of specimen there is a technique that is most suited, and therefore, if possible, this is the one that should be chosen. An additional point is that in one laboratory not *all* techniques should be in use, because otherwise the technical personnel do not develop sufficient expertise in each technique.

2.6 Cell Block Technique

The cell block technique takes an intermediate position between histological and cytological techniques. On the one hand, cells are embedded and cut, as in histological techniques; on the other hand, single cells and cell groupings, removed from the mother tissue, are used as in cytology. The cell sizes and configurations are as in histology because the cells are cut and do not spread over the glass slide as in other purely cytological techniques. The main advantage for the pathologist is that the cells resemble those seen in histology, and organoid patterns are clearly visible.

Two types of cytological material are well suited for the cell block technique: sputum and coelomic effusions.

2.6.1 Cell Block Technique for Sputum

The use of sputum for the cell block technique was described by Gray in 1964. It is thought to be preferable for the diagnosis of paracoccidiomycosis (Iwama de Mattos, 1979), but in our hands it has also proved highly effective in cancer diagnosis. The relative proportion of cancer cells on the slides

cytologic smear   cells block

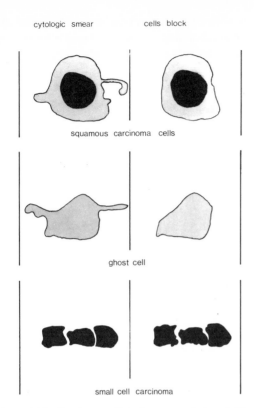

squamous carcinoma cells

ghost cell

small cell carcinoma

Figure 2.18 Carcinoma cells in the smear technique (left) and the cell block technique (right).

is larger, due to the fact that the benign squamous cells present in the sputum are visualised as thin strings and not, as in smears, as large polygonal structures (see figure 2.1). In contrast, the more spherical and thick cancer cells have the same configuration in the two methods, thus the *relative* proportion of these is larger in the cell block technique (see figure 2.18). In addition, immunoperoxidase stains can be performed on serial sections (Boon *et al.*, 1982*b*). From each block on three different levels, sections are cut. It requires much less time to screen the sections, in which the cells lie in concentrated areas, than to screen the six to eight smears made from the sputum. On these the cells are spread over the full surface of the slide.

The cell blocks can be stained with the differential Papanicolaou method, in which the abnormally highly keratinised cells stain orange–yellow. Highly keratinised squamous carcinoma cells are easily

detected with this method. The cytoplasm of large cell undifferentiated carcinoma and adenocarcinoma do *not* stain orange–yellow with this method (see appendix 2, Papanicolaou method 1).

The problem with cell-blocking sputum is that since it contains thick mucus, it takes approximately 24 h to fix it in formalin. Fixing in picric acid–alcohol takes 1–2 h. Formalin fixation can be shortened if the specimen, in a plastic container with formalin, is placed for 2 min in a microwave oven at maximal energy level (see section 10.2.3.3).

2.6.2 Cell Block Technique for Coelomic Effusions

As early as 1947, Chapman and Whalen reported a cell block technique for serous fluids. Some distinguished authors recommend the addition of anticoagulants to prevent clotting in serous fluid (Koss, 1979; Takahashi, 1981. However, in our laboratory we even encourage the fluid to clot (see section 10.2.7). The clot will contain a lot of cells, especially cancer cells when present, and is very good material for the cell block technique.

Since the organoid structure of the cell groupings is well visualised, the relationship between the fibrous core and the malignant cells can aid in the distinction between malignant mesothelioma and carcinoma.

2.7 Staining Results in Histological versus Cytological Specimens

Histological specimens consist of slices of cells of equal thickness; thus the cells in all parts of the slides are equally thick, and all the components of the cells are equally thick. In sharp contrast, a cytological specimen is very uneven: it consists of a mixture of single cells and much thicker cell groupings. In addition, the different parts of *one* cell vary in thickness. As will be discussed in great detail in the following chapters, the results of staining cells are dependent on the penetration of the dye in the cell, and its attachment to cell components. This attachment of the dye is fundamentally the same in cytological and histological specimens. However, the penetration in routine staining of one dye in even histological material is the same over the slide, whereas in uneven cytological material it differs a great deal between cells and between various parts of *one* cell. The results of this uneven staining of cytological material are well illustrated in the Atlas part of this book. In the discussion of the Papanicolaou and the Romanowsky–Giemsa staining methods, the implications of the unevenness of cytological preparations are further discussed.

In general, the staining times for cytological slides are shorter than for histological ones. Thus, when a histological staining technique is applied on cytological material, shorter staining times can be used (Sachdeva and Kline, 1981). In contrast, when a cytological staining technique is to be applied on histological material, longer staining times are needed. This is true for both the Papanicolaou and Romanowsky–Giemsa methods. In the latter, the desired 'Romanowsky effect' (purple staining of the nuclei) only occurs after prolonged staining times.

2.8 Resemblance of Cells in Histology and Cytology

It is clear that wet-fixed cells resemble most the formalin-fixed embedded and cut cells known in histology (table 2.2). This is true not only for cell

Table 2.2   Resemblance of cells in histology and cytology

	Wet-fix Papanicolaou method	Air-dry Romanowsky– Giemsa method
Nuclear dimension	Comparable	Larger, depending on cell type
Nucleus/cytoplasm ratio	Comparable in spherical cells	Different, depending on cell type
Chromatin pattern	Comparable	Different

sizes and nuclear/cytoplasmic ratios (see section 2.3.1), but also for chromatin patterns. The chromatin pattern evoked by wet fixation with ethyl alcohol resembles closely that of the formalin-fixed embedded cells (see Atlas, plate 15, and read chapter 4). In contrast, the chromatin pattern in air-dried cells differs significantly from these two (see chapter 4). This fact has proved to be of major importance in the history of cytology: generally speaking, in countries where the pioneers of cytology were haematologists, the air-dry Roman-owsky–Giemsa method is well developed and popular (The Netherlands, with Lopez Cardozo), but in countries where cytology was developed by, or in cooperation with, histologists and pathologists, the wet-fix Papanicolaou methods are thought to be superior (USA, with Papanicolaou, Koss, Reagan and Patten). As will become clear in the following chapters, both methods are highly valuable, if good care is taken in the slide preparation and staining technique.

Principles of Staining and Dyeing

3.1 Introduction

The cytologist is primarily concerned with the overall picture of the cell, showing nucleus and cytoplasm and, if possible, staining the nucleolus differentially. Often a distinction between nucleus and cytoplasm will be sufficient; in other cases a more detailed study of the cell's components is called for. So, in the first place, specific nuclear and cytoplasmic stains are needed (see chapters 5 and 6). Furthermore it may be important to study the specific constitution of the nucleus, i.e. its DNA and RNA content, or of the cytoplasm, i.e. protein, RNA, lipid and glycogen contents. In all cases it will be necessary to know something about stains and the mechanism of staining.

3.2 Staining

Staining is the process of colouring cells or components of cells or tissues which are otherwise colourless in light microscopy. Thus the goal of staining is visualisation by means of colouring.

3.2.1 Staining and Dyeing

In the context of this book, we will follow the common practice, and will use both the words 'staining' and 'dyeing'. However, we will use the term 'dyeing' exclusively if it covers our definition given in the following text.

Baker (1970) makes the following distinction between dyeing and staining. If an object is put into a dye solution and takes from it the dye molecule and not the solvent, the process is called dyeing. If, however, the subject takes up the solvent as well, this is called staining. Both processes result in a coloured object.

Most of the colouring processes using dyes stem from the textile industry. Bolles-Lee (1900) in his *Microtomist's Vademecum* refers nevertheless to the colouring of textiles as 'dyeing' and to that of biological material as 'staining'. We shall adhere to the following concept of staining: if a substrate becomes coloured by submerging it into a solution, the substrate is 'stained'; in this definition it is not formulated as to *how* the colouring is accomplished. If the colouring is caused by linking of a dye and a substrate component, we use the term dyeing; thus dyeing is then a special way of staining. In order to make this distinction, it is necessary to define quite clearly whether the colouring agent can qualify as a dye. We shall define dyes later (section 3.3).

Our definition of dyeing means that there are other ways of staining that do not use dyes. It is indeed possible to colour substrates in various other ways.

3.2.2 Other Ways of Colouring

The limited definition of staining implies the following requirement for the colouring agent (which can also be a non-dye): it must be soluble in a solvent and it must be able to colour a cell or tissue component. Under this definition, in addition to the techniques described in this book, fall enzyme staining techniques and immunotechniques. If they are to be applied to cytological material, the reader is advised to refer to books that deal with these techniques. In the opinion of the authors they do not at present belong to the 'routine cytological staining techniques' but might in the future come into this category. Some examples of staining mechanisms using non-dyes are given in section 3.4.

3.3 What Makes a Chemical Compound a Dye?

A dye is a coloured agent that is able to leave its solvent and to attach itself to a cell or a component of a cell or a tissue. If one were to list all dyes commonly used in microtechniques, it would at first glance be difficult to find a common characteristic. On closer examination it appears that certain groups of dyes share a common configuration in their molecules. It is to this configuration—*the chromophore*—that a dye owes its staining ability.

A compound lacking that chromophore, although otherwise closely related to a dye, will not be able to act as a dye. This will become clearer if one follows step by step the derivation of the triarylmethanes (to which the fuchsins belong), starting from a single methane molecule (figure 3.1). After replacing the hydrogen atoms by three aromatic rings and one hydroxyl group, the chromophore is introduced in stage V through hydration (stage IV will be discussed later). Finally, the compound is made into an active dye through salt formation with chloride (stage VI). In our example the chromophore is the quinoid ring. There are two other chromophore types: the $-N=N-$ configuration characteristic for the azo group, and NO_2, which is the chromophore of dyes of the nitro group.

From our definition of dyeing it becomes clear that the dye must be soluble and able to split in such a way that the part of the molecule that carries the chromophore can attach itself to the substrate. To this end an ionising agent, the *auxochrome*, must be added because the majority of dyes are aromatic salts which normally do not ionise. This is done in stage IV of our example, where the two NH_2 groups

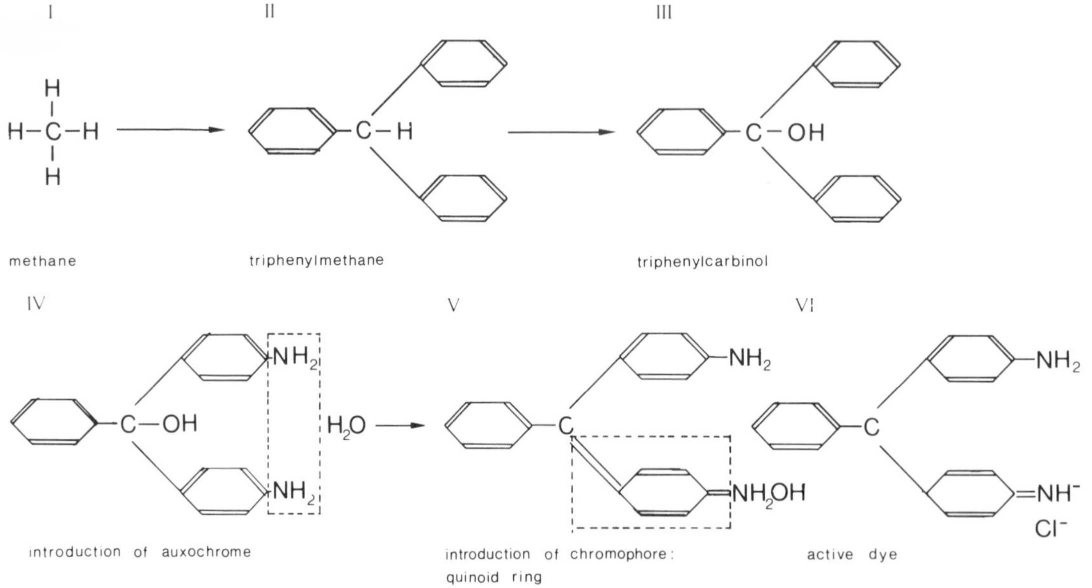

Figure 3.1   Derivation of triarylmethanes from a single methane molecule.

are added. Within each group (the quinoid, azo and nitro groups), basic as well as acid dyes can be found, depending on the auxochrome present. The effect of the chromophore on the 'dye molecule to be' is to change the energy level of its electrons so that its absorption maximum is shifted to the visible region of the spectrum. We shall not discuss this phenomenon any further but refer to Horobin's review in his book *Histochemistry* (Horobin, 1982).

Within each group, dyes can be found which for a long time have been called acid and basic. This is confusing because we have intentionally made the dye a salt. So what we should really say is that a dye is either anionic or cationic. In an *anionic* dye the chromophore is an anion. Its auxochrome can be either a sulphate or a carboxyl or hydroxyl group, and sodium is the usual cation with which to form a salt; anionic dyes can thus be represented by the general formula Na^+R^-. The acidity of the dye is commonly caused by the auxochrome SO_3^{2-}, but carboxyl and hydroxyl groups can also be found. A dye is called basic or *cationic* when the colouring is caused by the cation component of its molecule. In that case the anion of the salt is most commonly a chloride, although sulphate, nitrate, acetate and oxalate can also serve as anions. Basic dyes can thus be represented by the general formula R^+Cl^-. The basicity is caused by the auxochrome NH_3^-.

The chromophore can also carry both positive and negative charges. In that case we call the dye amphoteric. Whether it will react as an acid or a base will depend on the pH of the solvent. It will react as an acid if the pH is above its isoelectric point (IEP) (see section 3.12). Haematein and Light

Green are well known examples of amphoteric dyes. In our further discussion of dyes we shall retain the use of the well established terms acid and basic dyes, meaning anionic and cationic dyes, respectively.

It would also be possible to classify dyes according to the different atom or groups of atoms that can modify the molecule, for instance by replacement of the various hydrogen atoms in the aromatic ring. Modifications of this kind can also result in a change of colour. A good example of this is provided by the comparison of Eosin Y and Erythrosine B, both xanthenes (see figure 3.2). Eosin Y possesses three bromine atoms and dyes red. In the Erythrosine molecule the corresponding four hydrogen atoms are replaced by iodine and the dye stains blue. The difference in staining result is not caused by different attachment to the substance but by the formation of a bromide or an iodide, respectively; this causes a difference in effect on the energy level of the electrons within the molecule and thus on its absorption maximum, resulting in a different colour.

3.4 Classification of Dyes

A classification of dyes according to their chromophores is shown in table 3.1. It is obvious that the majority of dyes belong to the quinoid group. The table does not pretend to be complete. Only those dyes are included which are, for one reason or another, mentioned on these pages. The formulae and individual characteristics will be given in

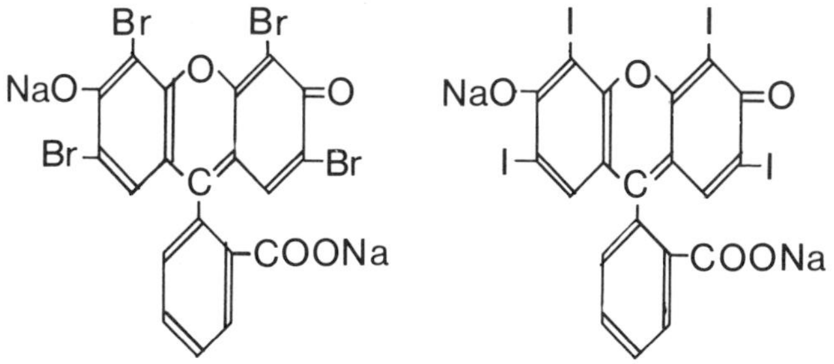

Figure 3.2 **The difference between Eosin Y (left) and Erythrosine B (right): the bromium of the former is replaced by iodine in the latter.**

Table 3.1   Classification of dyes

Chromophore	Basic chromophore	Use	Acid chromophore	Use
Quinoid ring Triarylmethanes	Basic Fuchsin	Schiff's reagent	Solochrome Cyanin R	Nuclear and cytoplasmic dye
	Pararosanilin	Nuclear dye		
	Rosanilin	Nuclear dye	Aniline Blue (amphoteric)	Cytoplasmic dye
	Methyl Violet	Nuclear dye	Acid Fuchsin	Cytoplasmic dye
	Crystal Violet	Nuclear dye and Gram stain		
	Methyl Green	DNA	Light Green Y (amphoteric)	Cytoplasmic dye
			Fast Green (amphoteric)	Cytoplasmic dye
			Coomassie Brilliant Blue	Cytoplasmic dye
			Haematein (amphoteric)	Nuclear dye
Haematein Anthraquinones	Alcian Blue	Mucopolysaccharides	Alizarin Blue	Nuclear dye
	Luxol Fast Blue	Fat	Purpurine	Demonstration of calcium
	Cuprolinic Blue	Nucleoli, RNA	Kernechtrot	Nuclear dye
			Carminic Acid	Nuclear dye
Xanthenes	Pyronin Y	RNA	Eosin Y	Cytoplasmic dye
	Rhodamine B	+ Nile Blue → Rhodanile Blue, nuclear and cytoplasmic dye	Erythrosine B	Cytoplasmic dye
			Gallein	Nuclear dye
	Acridine Orange	Nucleoli	Fluorone Black	Nuclear dye
			Methyl Fluorone Black	Nuclear dye
Azines and related dyes (quinone-imines) Azines	Neutral Red Rhodanile Blue	Vital dye Nuclear and cytoplasmic dye		
Oxazines	Nile Blue	Nuclear dye		

Group	Dye	Use
Thiazines	Gallamin Blue	Nuclear dye
	Coelestine Blue	Nuclear dye
	Gallocyanin	Nuclear dye
	Gallo Blue	Nuclear dye
	Cresyl Violet	Nuclear dye
	Brilliant Cresyl Blue	Vital dye
	Thionin	Nuclear dye
	Azure A,B,C	Nuclear dyes
	Methylene Blue	Nuclear dye
	Methylene Green	Nuclear dye
	Toluidine Blue	Nuclear dye
	Methylene Violet	Nuclear dye
Azo group, N=N Azines	Janus Green B	Vital stain
	Azocarmine	Nuclear dye in Heidenhain's Azan
Monoazo-	Janus Green B	Vital stain
	Orange G	Cytoplasmic dye
	Ponceau de Xylidine	Cytoplasmic dye
Diazo-	Bismarck Brown	Cytoplasmic dye
	Amido Black B	Nuclear dye (nucleoli)
	Pontacyl Dark Green	Nuclear dye (nucleoli)
	Oil Red O	Fat stain
	Biebrich Scarlet	Cytoplasmic dye
	Chlorazol Black E	Nuclear and cytoplasmic stain
Triazo-	Sudan Black	Fat stain
Nitro group, NO$_2$	Picric acid	Fixative and cytoplasmic dye
	Naphthol Yellow S	Protein dye
	Dinitrofluorobenzene	Protein dye

Appendix 4, together with references on their applications. The classification conforms mostly to that of Baker (1970), who groups together the various dyes according to their common chromophore. To this we have added a division according to the chemical reaction of the dyes, and therefore their use.

The amphoteric dyes can be further subdivided into acid dyes with weak or with strong active basic groups. Light Green forms a good example of the latter, as will be seen in section 6.3.2.2.

For practical purposes we shall divide the dyes into nuclear (staining nuclei), and cytoplasmic (staining cytoplasm). The nucleolus can be stained by both, depending on the staining procedure (see chapter 7). Each dye will be dealt with separately, and then its use in combination with others in polychrome stains will be described. The various staining methods in which the dye is used are presented in chapter 6. The neutral strains, staining both nucleus and cytoplasm, are discussed in chapter 9. Fat stains are included in our classification despite the fact that they do not qualify as dyes as far as ionisability is concerned.

The methods given are the results of the research and experience of many people. Some of the techniques have stood the test of time for almost a century. Successive authors may have advocated their own modifications, but the underlying principles still hold good and are now more fully understood.

3.5 Other Means of Colouring

Staining can also be achieved by applying agents which colour cells or tissue components but which do not qualify as dyes.

3.5.1 Staining with Non-dyes

Colouring cell components with compounds which are not dyes can be achieved in several ways, of which we shall discuss three:

(1) staining with colourless reagents that become proper dyes when attached to certain chemical compounds (Schiff's reagent);

(2) staining with coloured reagents that do not attach themselves to chemical compounds in the cell but dissolve in them (the fat stains);

(3) staining with metals, which either impregnate tissue and become coloured by certain tissue elements (as in the demonstration of nerve cells) or link with certain chemical compounds in the cell and become visible in electron microscopy.

3.5.2 Staining Methods using Schiff's Reagent

3.5.2.1 Introduction

The reactive compound is obtained by interaction of Basic Fuchsin (which consists mainly of pararosanilin) and sulphurous acid (H_2SO_3). By this reaction the dye is turned into a colourless compound by losing its chromophore (see figure 3.3). The chromophore is reintroduced by the reaction with aldehydes to which the reagent can link. The aldehydes can be present or formed:

(1) They can be present in the cell as free aldehydes (as in the demonstration of lipids).
(2) They can be introduced by oxidation of polysaccharides (in the determination of polysaccharides).
(3) They can be introduced by hydrolysis of the DNA–histone complex (in the determination of DNA).

3.5.2.2 Plasmal Reaction for the Detection of Lipids

Free aldehydes as in plasmalogens occur in the form of palmylaldehyde or stearaldehyde. Their reaction with Schiff's reagent is known as the plasmal reaction for the detection of lipids.

3.5.2.3 Periodic Acid Schiff (PAS) Reaction for the Detection of Polysaccharides

In this method, the 1,2-glycerol groups on the polysaccharide chains are oxidised by periodic acid (H_5IO_6), which introduces free aldehyde groups ready to react with Schiff's reagent. This reaction is used for demonstration of the polysaccharides glycogen and mucin (figure 3.4).

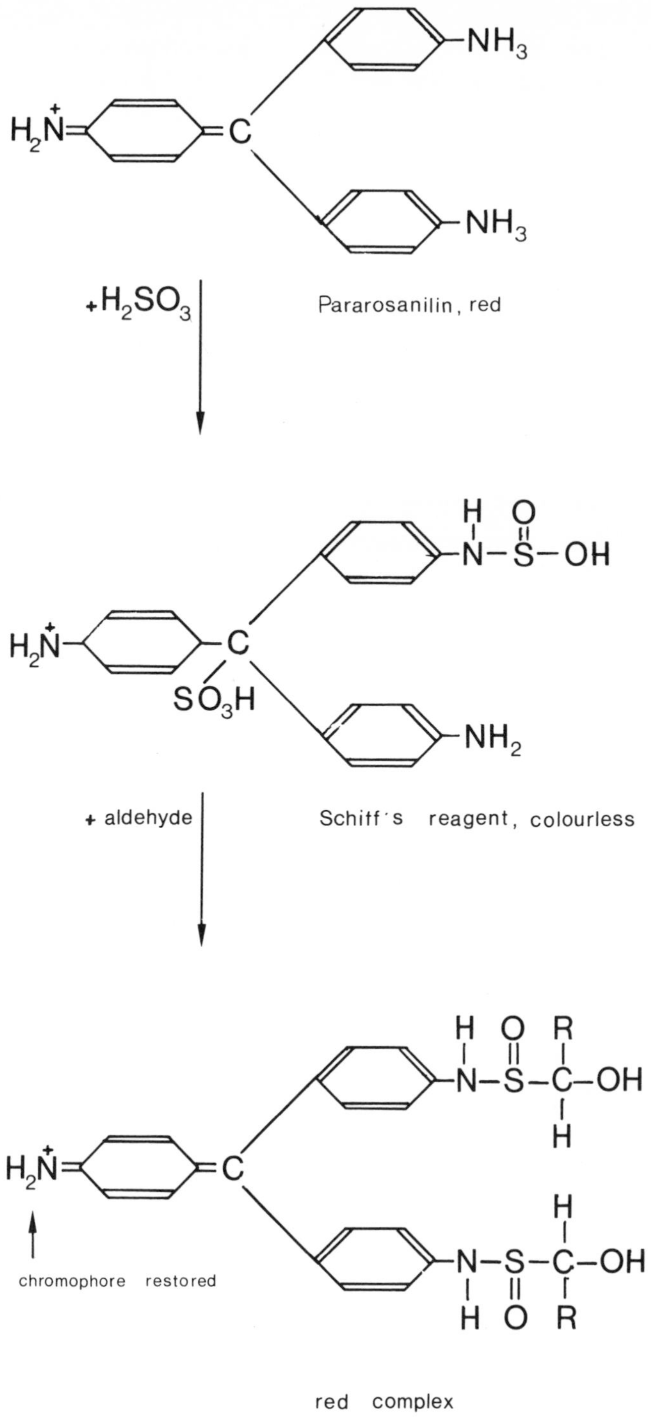

Figure 3.3 The formation of Schiff's reagent and its reaction with aldehyde.

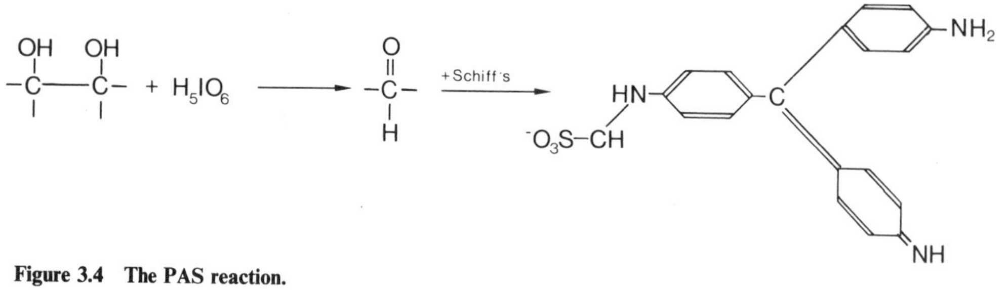

Figure 3.4   The PAS reaction.

3.5.2.4 Feulgen Reaction for the Demonstration of DNA

In this method the aldehydes in the DNA–histone complex of chromatin are freed by hydrolysis. The mild hydrolysis by hydrochloric acid removes RNA and leaves DNA in its place. It splits off adenine and thymine from deoxyribose. Thus the aldehyde groups of deoxyribose are freed to react with Schiff's reagent. The method provides a quantitative assessment of the cell's DNA content, because it stains only DNA and not, as for example, the haematoxylins do, the nucleoproteins (figure 3.5).

3.5.3 Fat Stains

Non-polar lipids cannot be coloured by proper dyes. Instead, coloured compounds can be used which are also not ionised but which carry a chromophore. Such compounds, named 'lysochromes' by Baker, must have some reason to 'stay' with the lipids and not in the solvent in which they are presented to the lipids. In other words, they must be soluble in lipids than in their solvents. This is the case for coloured agents like Sudan III, Sudan Black and Nile Red. There is thus no linking involved in this kind of staining (neither agent nor substrate are ionised); the colouring is based on a difference in solubility.

3.5.4 Staining with Metal Compounds: Osmium Tetroxide, Uranyl Acetate, Lead Salts

There are various staining methods in which the deposition of metal compounds in specific cell components plays a vital role, either to visualise specialised cell components, or to increase contrasts in electron microscopy. The deposition of reduced

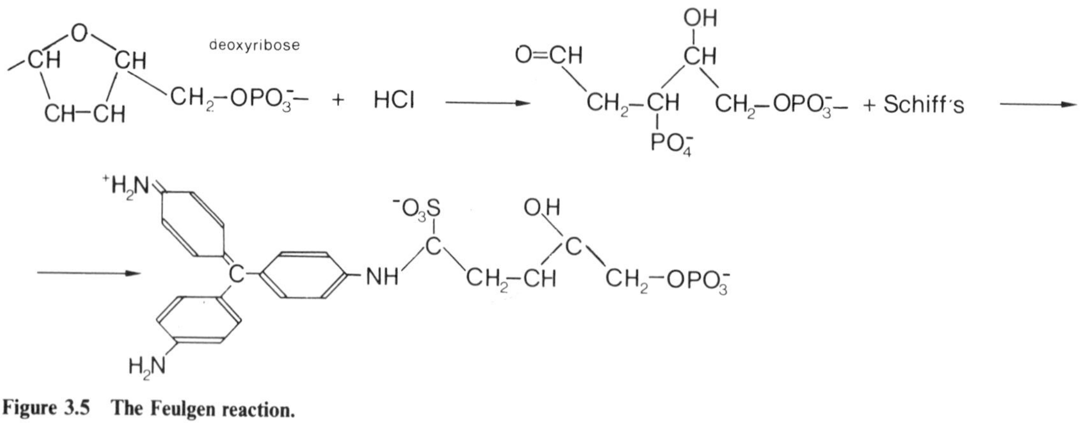

Figure 3.5   The Feulgen reaction.

silver is used to trace axons and dendrites of nerve cells. Tissue is impregnated with silver, which is reduced in the axons and dendrites (probably at sulphydryl groups) and thus visualised. The process is very much like the developing of a photographic film. There is no indication that the silver particles link with any tissue component.

Staining for electron microscopy also makes use of metals. Osmium tetroxide not only fixes cells but its presence can actually be seen, especially at lipid-rich sites like membranes. Unsaturated lipids can reduce osmium tetroxide to a black compound which can also be seen in light microscopy. Heavy metals are used to increase the electron-scattering power of certain cell components. Uranyl acetate combines with lipids so that these become more prominent in electron micrographs.

3.6 Staining with Natural Dyes: Past and Present

Of the dyes mentioned in table 3.1, some are natural but the majority are synthetic. This does not necessarily mean that natural dyes are less important. On the contrary, one of the most widely used dyes in microtechnique is a natural dye, namely Haematoxylin.

The sources of natural dyes are numerous, and the dyes serve a variety of purposes. Port wine can be used to colour marine animals, and dried bodies of female insects can be used to stain nuclei. An extract of red cabbage can also be used for the latter purpose. Virtually all parts of plants can supply some dye or other and have been used for that purpose from ancient times. The roots of madder were used in the Middle East to dye goatskins, and the fact that bones become coloured if an animal eats madder was known by the ancient Chinese. The wood of brazilwood from which the dye Brazilin can be extracted was imported from Indonesia into Europe in the 12th century. It resembles Haematoxylin, which is also extracted from wood. Anthocyanins, which can be extracted from a wide variety of flowers, have had many applications in zoological and botanical staining. Fruits and berries (as well as their skins—hence the usefulness

of port wine) are other good suppliers of dyes. Our famous compatriot van Leeuwenhoek is reported to have used an extract from the stigmas of crocuses to stain muscle fibres (van Leeuwenhoek, 1674, 1679, 1702), and there are many more examples throughout history.

The majority of natural dyes can now be made synthetically. Orcein, which was originally extracted from lichens, Alizarin, extracted from the madder plant, and Indigo, from wood, are well known old dyes which can now be manmade. But some natural dyes remain in use and of these Haematoxylin still ranks as the most important nuclear stain in routine histology and cytology. Because of this we shall deal with it separately in section 5.2. Cochineal is another natural dye which is still used as such. It is made by grinding the bodies of the female cochineal insect. The deep red dye was used by the Aztecs and came to Europe after the conquest of Mexico in 1520. Cochineal itself is not much used as a dye but its derivative, carminic acid, and the aluminium lake of the acid, known as carmine, are still in use. The dye finds its most important application in Best's Carmine stain for glycogen to which it attaches itself by hydrogen bonding. There is such a wealth of dyes in nature that it might well be that some very good dye has escaped the attention of dyers and it might be worth while, time permitting, testing them on cytological material. For an extensive review of the natural dyes see Lillie (1977).

Much of what we now know about dyeing mechanisms comes directly from studies of the dyeing process in the textile industry using natural dyes. This has a very long history, which makes interesting reading. For those interested we recommend the books by Vickerstaff (1950), Bird (1951) and Venkataraman (1952).

3.7 Mechanism of Dyeing

3.7.1 Introduction

In chapter 1 we have summarised and discussed the main chemical compounds in the nucleus and cytoplasm. Their structures are complex and so are their spatial configurations. In sections 3.3 and 3.4 we have looked at the structure of dyes and how they vary in size and chemical properties. We shall now discuss the various theories on the mechanism of staining. First, however, it is necessary to look at the ways in which chemical compounds can link together and how these means of attachment can be applied to the linking mechanism of dye and substrate.

3.7.2 Forces Involved in Linking Chemical Compounds

When ions of opposite charge are brought together, they combine to form a salt through so-called *coulombic or ionic bonds*. From the classification of dyes, we have already seen that they can be divided into acid and basic dyes. Yet all dyes are salts. In water they dissociate into their constituent ions. For the 'acid' dye, the anion is the chromophore, and for the 'basic' dye, the chromophore is the cation. Since the various cell components also contain anions, these can link to the chromophore cations, and the cations in the cell can be attached to chromophore anions. In both instances a salt is formed. Coulombic bonds are not very strong and can be broken down easily by water dipoles. For ionisation of dyes their solvents need to contain at least some water. This is always the case, although it is sometimes present in only very small quantities (for instance, in alcoholic solutions).

If two atoms share a proton (H^+), this results in a link called a *hydrogen* bond. This is, for instance, the case if oxygen or nitrogen, which have a strong negative charge, are bound covalently to hydrogen. The shared electron of this covalent bond will be situated nearer to the oxygen (or the nitrogen) atom, leaving the hydrogen atom with a positive charge. This gives the hydrogen atom the opportunity to attract another atom with a negative charge, which can be an additional oxygen or nitrogen atom. Amino (NH_3), carbonyl (C) or hydroxyl (OH) groups are frequently involved in the hydrogen bonding that occurs in staining. Water itself is a strong hydrogen bonder. Because of this, it is unlikely that staining processes occur-

ring in water involve hydrogen bonding, but it can be an important mechanism in staining in alcoholic solutions.

A strong link is formed through *covalent bonding*. Here two atoms share one or more electron pairs. They form the most common bonds in chemistry. The periodic acid–Schiff methods to demonstrate glycogen and the Feulgen reaction to locate DNA are good examples of this mechanism. If the shared electron pair belongs to only one of the atoms involved (provided by the so-called donor molecule or ligand), the bonding is called *dative bonding*. If the ligand has two or more donor atoms to offer, it is called a *chelate*. Usually these chelates are metals, and here we come to the often used *mordant dyes*. Very often the dye itself will not form a lasting strong link with the substrate but introduction of a metal will form a strong dye–metal–substrate complex through covalent (dative) bonding. We shall come back to mordant dyeing in a later section (3.9.2).

Molecules which are neutral but contain dipoles, or in which dipoles can be induced, attract each other by forces called van der Waals forces. They are only of importance when the molecules are close together. To be able to exercise van der Waals forces, molecules must be polarisable and large. The larger the molecule the stronger the van der Waals forces that it can exercise. Amines, nitro groups and sulphonic acids are groups which have large dipoles and thus can readily be involved in van der Waals attractions, which can occur in all kinds of solvents.

So far we have discussed links in which electric charges are involved. If a molecule has large or long uncharged groups, as in fats and carbohydrates, these hydrophobic groups tend to aggregate in aqueous solutions. This tendency of non-polar groups to aggregate in water is called hydrophobic bonding. Both dye and substrate can possess such non-polar groups, and when these are brought together, this can result in a hydrophobic dye–substrate bond. The energy required for this type of bonding is produced by the disturbance of the ordered state of the water. It is clear that hydrophobic bonding can *only* occur in water.

Some molecules are hydrophobic and others

hydrophilic, and some have hydrophobic and hydrophilic groups next to each other. Hansch π values are measures by which the hydrophobic—hydrophilic character of a compound can be determined. These values are negative for hydrophilic groups and positive for hydrophobic groups. Addition of the Hansch π values of the groups making up the molecule will tell whether a molecule is hydrophobic or not (Horobin, 1980). If there are ionic groups in the molecule, as there are in acid and basic dyes, the Hansch π value will be negative or zero and the compound will be soluble in water. If there are no ionic groups, as for instance in fat, the Hansch π value will be positive and the compound will be hydrophobic.

In this review we have only skimmed the surface of the theoretical background of linking forces. More detailed information can be found in Horobin's handbook on histochemistry, which includes a number of models for dye–substrate linking under various circumstances (Horobin, 1982), and in Zanker's paper on the principles of dye–substrate relations (Zanker, 1981).

3.7.3 Theories on the Mechanism of Dyeing

The links between dye and substrate that we have described in the previous section may individually be too weak. The actual link causing the colouring may be the result of a *combination* of attracting forces. In addition, physical properties of the molecules of both substrate and dye can contribute to linking.

Throughout the history of dyeing there have been opposing views on its mechanism. As we have seen in section 3.3, dyes can be acid, amphoteric or basic in their chemical reaction. The two main components of the cell, the nucleus and the cytoplasm, also differ in their chemical reaction (see sections 1.1 and 1.2), and thus it is to be anticipated that the nucleus is stained by basic dyes and the cytoplasm by acid dyes. A simple statement like this assumes an exclusive chemical explanation of staining due to ionic bonding: the positive charge of the dye bonds with the negative charge of the cell components and vice versa. This assumption is

certainly not shared by authors such as von Möllendorff (1924). He explains all dyeing processes as physical ones, governed by the differences in penetration of the dyes in the various tissue components.

Singer (1952) tackled the problem from a completely different angle: according to him the reaction between dye and tissue is always an electrostatic one. He even explained diffusion and penetration of dyes into tissue, which play such an important part in von Möllendorff's theory, by coulombic (ionic) bonding. From his experiments it became clear, however, that salt formation, in which one basic group would be expected to react with one acid group, could not always explain the many aspects of the staining process. There have to be other bonding forces as well, and therefore Singer had to admit that van der Waals forces, resonance, and covalent bonds were also involved.

Between these two extremes stand the views of Baker (1970) and Lillie (1977). Both authors take a much broader view of the whole staining mechanism and they include both physical and chemical forces of bonding.

The various ways in which chemical compounds can link together are governed by certain dye characteristics, some of which can be measured. For instance, the hydrophobic–hydrophilic character of dye molecules can be calculated, as has been done by Hansch *et al.* (1973). Whether a molecule is polarisable is determined by its number of covalent bonds which are involved in aromatic or conjugated groups and which can be counted in the structure formula. This measure, called the conjugated bond number, gives some idea as to whether van der Waals forces can be expected to be involved. The overall electric charge of the dye molecule will tell if coulombic bonding can take place. The size of the dye molecule or ion will have its influence on rate of penetration and also on the possibility of van der Waals forces occurring.

The relation between the structural features of dyes and the possible dye–substrate link is summarised in table 3.2. It is possible to list all these characteristics of dyes and thus get some idea as to which mechanism is involved in the staining process. In our description of dyes which are

Table 3.2   Relation between the structural features of dyes and the possible dye–substrate link

Structural feature	Determining factor for:
Overall electric charge	Presence of coulombic (ionic) bonding
Size, measured by ionic weight	Rate of diffusion into substrate; strength of van der Waals forces
Conjugated bond number	Presence and strength of van der Waals forces
Hansch π value as measure for hydrophobic/hydrophilic character	Presence of hydrophobic bonding
Presence of suitable groups, e.g. NH_3, OH, CO	Possibility of hydrogen bonding
Presence of donor atoms usually oxygen or nitrogen at suitable sites	Covalent bonding; possibility of mordanting

commonly used in cytology (appendix IV]. We provide the Hansch π values for most dyes. For more information on Hansch π values of molecular fragments we refer readers to Hansch's original work (Hansch *et al.*, 1973). The importance of such a list of data lies not only in the possibility of gaining some insight into the staining mechanism but also in the possibility of finding a replacing dye if needed. If, for some reason or other, one needs or wants to use a different dye with similar staining results, one can choose a dye with similar characteristics from this list.

From this discussion it will have become clear that it is impossible to draw up one single model, as some authors have tried, to explain the mechanism of staining. What actually happens during staining is determined by the physical and chemical properties of the dye and substrate involved. Hence the more we know about them, the closer we can come to understanding the various staining mechanisms, and the better we can manipulate staining to suit our specific needs.

3.8 Dyeing in Cytology

When trying to explain the various aspects of dyeing in cytology, it seems sensible to take both physical and chemical properties of dye and cell component into account. The rate of penetration, on which differential staining in histology is often based, is also important in cytological staining. Yet since cytology specimens are uneven, the results differ. In sections the difference in the rate of penetration in the various tissue components is responsible for the differential staining results. In cytology there is a difference in penetration of the dyes in single cells and cell groupings, and thus the staining results differ between cells on the one hand, and between single cells and cell groupings on the other. Horobin's affinity models for biological staining concern dye–tissue relations and are not necessarily totally valid in cytological staining (Horobin, 1982).

3.9 Methods of Dyeing

3.9.1 Direct Dyeing

When a dye is used as a direct dye not needing a mordant, we are concerned with direct dyeing. Eosin Y is a good example of a direct dye, and so are the thiazine dyes (see section 5.3).

3.9.2 Mordant Dyeing

Not all dyes are suitable to be used as direct dyes. This is especially so for dyes that are amphoteric.

The bond between ion and cell component to be dyed would in such a situation not be strong enough because part of the dye molecule has a charge opposing that of the cell component. When using such dyes, an extra bridging ion is introduced which positions itself between the dye ion and the cell component, thus forming a strong link. The resulting strong linking of dye–mordant–cell is called a lake (see figure 3.6).

Dye and substrate molecules that are suitable for lake formation must have electron donor atoms. In nuclear mordant staining, the nucleic acid phosphate groups and the carboxyl groups of proteins are especially suitable for lake formation. Iron, aluminium and chromium are most commonly used to build the necessary bridge. For dyes, oxygen and nitrogen are suitable electron donors.

Not every dye is suitable to form a lake. It needs to have a special configuration in its molecule. The combination of a double-bonded oxygen atom and an OH group next to it often provides a specially suitable place for the metal atom to settle. The metal atom then becomes 'gripped' between the oxygen atom and the OH group. Such a pincer-like grip is called a chelate. This kind of staining is nearly always used regressively (see section 3.10.2). This means that the excess amount of dye which is weakly bound to other cell components is removed afterwards. This removal of excess dye is called differentiation, and is the result of breaking up the weaker complexes of tissue–mordant–dye. Thus the excess dye is set free. This is usually achieved by using excess acid.

It is also possible to remove excess dye molecules by providing more mordant molecules than are needed for the sites available in the substrate component. In that model, excess dye molecules are 'taken away' by the mordant. This is for instance the case for Weigert's Haematoxylin. In that method differentiation takes place during immersion in the actual mordant solution after staining with Haematoxylin. The metal ions used for mordanting not only establish the link between dye and substrate but also have an effect on the resulting colour of the dye. This can be explained by a change in the energy level within the dye molecule, result-

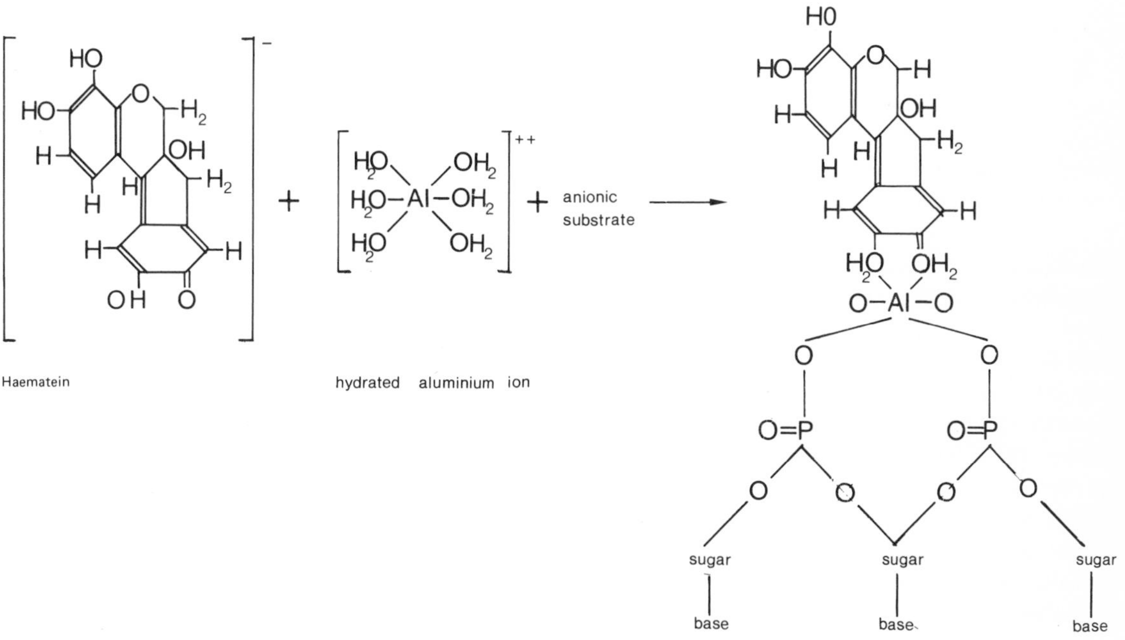

Figure 3.6 Formation of the lake between Haematein, aluminium and nucleic acid.

ing in a shift in its absorption maximum. This becomes very obvious when comparing the staining effect of iron Haematoxylin (like Heidenhain's and Weigert's) and aluminium Haematoxylin (like Delafield's, Ehrlich's, Harris's, Mayer's or Cole's). Nuclei stained with iron as a mordant become black whereas staining with aluminium as a mordant gives blue nuclei. For details of the composition of these Haematoxylins, see table 5.1.

Other dyes suitable for mordant dyeing apart from those already mentioned are Gallo Blue, Solochrome Cyanin R, Gallocyanin and Coelestine Blue. They are all used as nuclear dyes.

3.9.3 Orthochromatic and Metachromatic Staining

In appendix 4 we give details of those dyes generally used in cytology. One of its characteristics is the dye's spectral absorption maximum or maxima. With an absorption maximum of, for instance, 515–518 nm, the dye will be red. If the colour of the substrate dyed also has its absorption peak in that same region (and thus has the same colour), the dyeing process is called *orthochromatic*. If a dye in solution has a certain colour but stains the cell component with another colour, due to a change of physical but not of chemical properties of the dye, the process is termed *metachromasia*. The change in absorption peak here is caused by a chemical compound in the substrate component changing the monomer form of the dye to a dimer. This makes the substrate component a *chromotrope*. A prerequisite for a substrate component to be a chromotrope is the presence of the sulphuric, phosphoric or carboxyl radical. The extent of the metachromatic effect of these groups decreases in that order. Of the three the phosphoric group is the one which is of the greatest interest to the cytologist because the polymerisation capacity of DNA, used in the Romanowsky–Giemsa method, is based on it. It is not only the nature of the acid group which determines the degree of the chromotropic effect, but also the frequency with which it occurs on the substrate. Hence highly polymeric substances with many suitable acid groups will be more chromotropic than those consisting of more simple

molecules with fewer suitable acid groups available.

Another example of a commonly used metachromatic dye is Eosin Y: the monomer is red and the dimer orange-red (Marshall *et al.*, 1981).

At room temperature an aqueous dye solution of low ionic strength will often contain the dye predominantly in its monomer state. When the concentration is increased and with it its ionic strength, aggregation is increased and many more dye ions cling together forming dye micelles. Precipitation of the dye is the extreme form of aggregating dye molecules. This grouping together of the dye ions is the cause of a shift in the absorption peak of the dye solution. Both cationic and anionic dyes can aggregate, but the former are best known to the cytologist.

The triarylmethane Methyl Violet and the oxazine Coelestine Blue can dye metachromatically. However, the majority of metachromatic dyes are to be found among the thiazines: Methylene

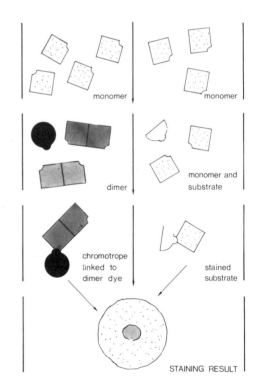

Figure 3.7   **Left: metachromatic staining by the dimer configuration of a dye. Right: orthochromatic staining by the monomer configuration.**

Blue, the Azures and Toluidine Blue. In all these instances the absorption peak of the dye solution is shifted from blue to red in circumstances favouring aggregation. It is believed that van der Waals forces play a major part in the attraction of the dye molecules. The degree of aggregation and hence metachromasia is not only dependent on dye concentration and temperature but is also strongly influenced by the nature of the solvent. Addition of salt to an aqueous solution increases the metachromatic effect; addition of methanol, ethanol or dimethylsulphoxide decreases it. If the dye stains the cytoplasm orthochromatically and the nucleus metachromatically, a differential staining effect is achieved: figure 3.7 (see also chapter 9).

3.10 Progressive and Regressive Staining

3.10.1 Progressive Staining

When a substrate is brought into contact with a dye solution, the dye ions diffuse into the substrate and (first) link with the most suitable substrate components. The dye–component relationship is deter-mined by the nature of the solvent, the dye concentration, the surface area of the component, and the number of suitable binding sites available. The diffusion rate is also one of the determining factors of the rate of staining of a given component. After exposure to the dye solution for a certain length of time, staining equilibrium is reached (see figure 3.8). At that point no more dye ions can be linked to the substrate component. If the staining time is shorter than 'equilibrium time', only fast-staining components of the substrate will be stained. The result is then differential staining by making use of the difference in staining rate. This type of staining is described as progressive. With this staining method best results are obtained with large or aggregated dye ions, which diffuse slowly.

Different components of the substrate have different staining rates, and thus different equilibrium times. Equilibrium time depends on the affinity of the component for the dye, on the one hand, and on the possibilities of the dye to reach the component, on the other. In all progressive staining methods, precise staining times are very important in order to have the desired staining effect.

Progressive staining is possible not only with one

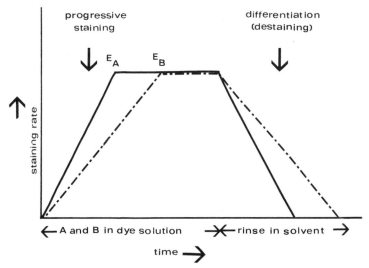

Figure 3.8 Sequence of a staining procedure schematically represented for a substrate containing two substances, A and B, which have different staining rates. (After Horobin, 1982.) ——, staining rate of substance A;–·–·–·–, staining rate of substance B. E_A, equilibrium point for substance A; E_B, equilibrium point for substance B.

dye but also with a combination of dyes (e.g. the cytoplasmic stain in Papanicolaou's staining method). It is very important with such methods to observe the standardised staining times. Longer staining times lead to shifts in staining patterns (see figure 4.3).

3.10.2 Regressive Staining

The staining method in which the substrate is first overstained and consequently partly destained is called 'regressive staining'. After a prolonged contact between substrate and dye, less strong links between other components will also occur, or the dye will simply be deposited in the substrate without any lasting links between substrate components and dye. There then exists no colour difference between the different components, and everything will be coloured the same. This situation is called 'overstaining'. A certain degree of destaining is needed to visualise the different components. The components which have only very weak links with the dye will destain first, by which means the other components will become visible. This process is called 'differentiation', and can be achieved in several ways. Sometimes a rinse in water is sufficient to remove excess dyes from these weakly linked sites. This is more the case for acid dyes than for basic dyes. The latter are more easily removed by the alcohols which are used for dehydration. Of these methanol has the strongest effect and tertiary butanol the weakest. An acid dye can be removed by a base and a basic dye by an acid. The isoelectric point of the substrate components plays an important part in this.

In regressive staining methods the staining times must be longer than the equilibrium time of the slowest staining component (see figure 3.8). The time of differentiation is extremely important. Too long differentiation times will lead to destaining of the component one wants to visualise (as is the case for chromatin). Too short differentiation times will lead to colouring components one does not wish to stain, for instance regions of the cytoplasm. (See figure 3.8.)

Mordant dyeing is nearly always done regressive-ly and in the case of nuclear dyeing the destaining agent is an acid. More about this aspect can be found in section 3.9.2.

Higher dye concentrations give a higher diffusion rate. Regressive staining therefore usually means working with higher dye concentrations than when staining progressively. Table 5.1 shows clearly that dye concentration determines whether Haematoxylin is used progressively or regressively. Because of its widespread use, much research has been done on this dye and we shall discuss this aspect in more detail in chapter 5.

3.11 Differential Dyeing

The fact that some substrates take up a particular dye and others do not is used to differentiate between tissue and cell components. If more than one dye is used, different components will be coloured differently (differential dyeing). This goal can be achieved in various ways.

3.11.1 Differential Dyeing Using More than One Dye

The preference of a substrate for a certain dye or vice versa may be due purely to the chemical properties of both. For dyes we have already summarised these properties: net electric charge, polarisability, hydrophobic–hydrophilic character, and size of molecule. They all contribute to the affinity of a dye for a certain substrate.

Whether a substrate will accept a dye depends on the presence, availability and character of bonding sites. Suitable sites may not be available because the substrate through which the dye has to travel may be too dense. This difference in density is a very important factor in the Papanicolaou staining method, for instance, as we shall see later. Another reason why a substrate cannot accept all dyes may be the fact that the suitable sites are covered by other substrate molecules. Fixation plays an important part in the 'unveiling' of these sites by breaking up protein linkages, i.e. denaturing proteins or links such as those between nucleoproteins and nucleic acids.

Assuming that all reactive sites are available, the substrate may or may not accept a dye; in other words only certain elements of the substrate may become stained. This will provide a certain degree of differentiation between them. This may not be enough for a critical study of the substrate and one may want to add another colour, leaving the first one in situ. The combination Haematoxylin–Eosin is a well known example in histology.

Another possibility is to present a substrate with several dyes at the same time, which is the case in a number of frequently used staining methods in histology and cytology: Masson's trichrome stain, Papanicolaou's polychrome stain, the Methyl Green–Pyronin Y combination, and the blood stains, to mention just a few. The combination of dyes may consist of those that link to only one particular component of the substrate, which is for instance the case for nuclear staining with Methyl Green–Pyronin Y: DNA takes up Methyl Green and RNA takes up Pyronin Y. In such a combination, competition between dyes is irrelevant. The same is almost true for the Haematoxylin (nuclear) and Eosin (cytoplasmic) stain. A different situation arises if the substrate is confronted with a combination of dyes which would, when used individually, each stain some or all of its elements. In that case the differential staining of substrate components must be due to some kind of competition. Papanicolaou's combination of anionic dyes gives a very good example of this. Here Eosin Y (with four aromatic rings in its dye ion) and Light Green (with five aromatic rings) are offered simultaneously to cell samples. If the sample contains blood cells, the erythrocytes will be dyed red by the smaller dye ion, Eosin. Erythrocytes have a dense structure. If the even smaller dye ions of Orange G (with three aromatic rings) are added to the Papanicolaou stain, there is an added competition between Eosin and Orange G (see section 6.3.2.2). Keratin molecules have a very dense structure owing to many disulphide bridges (see section 1.2.1.1), and the more keratinised the cell is, the denser the structure. The competition is clearly won in the densest structures by the smaller dye ion.

If the rate of penetration is the determining factor for successful differential staining, it is important to adhere to strictly limited and controlled staining times. This is certainly so for Papanicolaou's method where differentiation changes over time.

Differential dyeing can also, of course, be achieved by making use of the ortho- and metachromatic staining of one dye acting as if there were two dyes (see section 3.9.3).

3.12 Influence of pH on Dyeing Proteins

From the behaviour of proteins in solutions of varying pH (see section 1.2.1.1), it is obvious that the pH of the staining solution has a great influence on the staining result. This is especially so if the staining mechanism operates by salt-linking between dye and substrate ion.

The charged groups of cell proteins that form salts with basic dyes are the free carboxyl groups of aspartic, glutamic and hydroglutamic acids, and the free acidic groups of phosphoproteins and mucoproteins. The charged groups of cell proteins that form salts with acid dyes are the free basic groups of the amino acids lysine, histidine and arginine. The extremes of pH at which these groups are maximally dissociated are respectively 2 and 11 for the free amino and carboxyl groups (Singer, 1952). It is important to realise that in protein staining (Orange G, Eosin Y, Light Green), when applied under proper conditions, the presence of a particular amino acid or group of amino acids is demonstrated, and not a specific protein (Singer, 1952).

Figure 3.9 shows the character of acid, basic and amphoteric dyes at various pH. Acid dyes react as acids throughout the range of pH which is normally used (2–9), and basic dyes react as bases within that same range. The amphoteric dyes react as acids at pH values below their isoelectric point and as bases above it (Baker, 1970). The intensity of the colour in protein staining is thus dependent on the pH of the staining solution. This is illustrated in the graphs of the changes of staining intensities at various pH values in chapter 6 (figure 6.4). Note that the acid dye Eosin does not have its peak at a low pH, as most acid dyes do. Most of these stain

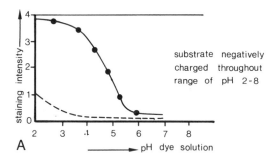

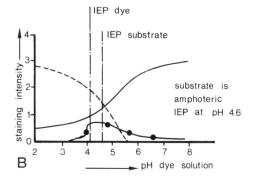

Figure 3.9 Relationship between intensity of dyeing by an acid dye (– – –), basic dye (——), and an amphoteric dye (– ● –) of (A) an amphoteric substrate, and (B) an acid substrate (after Baker, 1970).

best in an acid solution, and most basic dyes in a basic one: however, there are many exceptions to this rule. Also adding salts ('salting') can have a profound effect (Singer, 1952).

The pH dependence of dyeing is discussed at length by Singer (1952). Figure 3.9 shows the relationship between the intensity of staining of an amphoteric protein by an acid, amphoteric and basic dye, respectively. The graph is obtained from dyeing a film of the acid protein collodion. It illustrates that the curves for the acid and basic dyes meet each other roughly at the IEP of the protein. This phenomenon is used by some authors as a means of measurement of the IEP of proteins. By so doing it is assumed that only ionic bonding is involved in such staining. That this is not always the case is recognised by Singer, and further emphasised by Horobin and Bennion (1973), Zanker (1981) and Horobin (1982).

When other forces provide the bond combining

dye and protein, the electrostatic forces are nevertheless still important by attracting the dye to the protein within the range of these other forces.

Great caution is mandatory when one tries to explain the effect of dyeing proteins in cells and tissues by applying the data obtained from staining proteins in solution with acid and basic dyes. In cells and tissues, the dye is bound to three-dimensional solid protein, which might react quite differently. In addition, the internal pH of the cell might play a decisive role (Horobin, 1982).

Fixation has a profound influence on the staining of proteins. Unfixed proteins have little capacity to bind dye. In general, a noticeable increase in stainability with both acid and basic dyes occurs after fixation. However, this increase might differ between the two. This effect varies between the various fixatives. For instance, formaldehyde favours binding with acid dyes (Singer, 1952). Last but not least, fixation can change the isoelectric point of proteins: formaldehyde lowers it, but ethyl alcohol has no effect.

The importance of pH for staining solutions becomes particularly clear in those methods which are aimed at quantitation of proteins. Here we have only summarised some methods for the dyeing of nuclear components in relation to the pH of the dye solution. Fast Green FCF is a dye which is used to distinguish between basic (histone) nuclear proteins and acid (non-histone) proteins (Alfert and Geschwind, 1952; Dhar and Shah, 1982). In the same way, the acid dye Biebrich Scarlet in alkaline solution is used to identify basic proteins according to their degree of basicity. For a more detailed discussion of protein staining in quantitative cytochemistry, see Tas *et al.* (1980).

The importance of pH in the routine staining methods of Papanicolaou and Romanowsky–Giemsa is discussed in chapters 6 and 9, respectively. We are not able to explain the effects of pH changes in these two methods to our full satisfaction, even when the points listed above are taken into account. Moreover, in the Papanicolaou method an additional but decisive factor is the difference in penetrating dense structures by the competing acid dyes Eosin, Light Green and Orange G.

4

Fixation

4.1 Introduction

So far we have looked at the presence and structure
of cell components (chapter 1), the structure of
colouring agents (sections 3.3 and 4) and the ways
of linking the two (section 3.7). If staining, and thus
visualising, cells and their components could be
based on the simple linking of colouring agents to
cell components in cells that in all aspects are
identical to those in the original sample, then it
would be reasonable to assume that all cell components would still be present, that the colouring
agents would be able to enter the cell, and that they
could easily reach those cell components with
which they are best suited to form a link. As far as
the last of these is concerned, it would be necessary
for the substrate not to be obscured by any other
cell component. That this is so in the living cell has
been shown for chromatin and proteins in section
1.1.1, and sections 1.2.1.2 and 3. Also, we seem to
be taking for granted that the cell components that
we want to visualise by colouring are all still there,
and all in their original positions, at the moment of
staining. That these assumptions are not true will
become clear in this chapter.

Ranvier stated that *fixation is necessary* to stain
cells (1895). However, this is not true: it is possible
to stain living cells by the so-called vital staining
methods, as described in chapter 8. The stained
unfixed cells cannot be kept, and so, in diagnostic
cytology, methods using fixation of cells are far
more commonly used. Without any kind of
preservative treatment or fixation, cells deteriorate

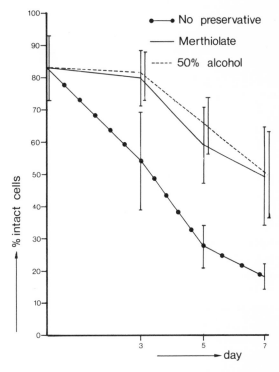

Figure 4.1 **Degeneration of cells in urine due to the presence
of bacteria. Within 3 days, 50% of the cells are no
longer intact if no bacteriostatic or alcohol (pre-
fixation) is added to the urine.**

by autolysis (self-digestion) and by the action of
bacteria and moulds (see figure 4.1).

In order to succeed in obtaining permanent,
stained slides, there are a number of measures we
have to take before we can undertake the process of
staining. These tasks are left to agents which we call

fixatives. In addition, fixation can be achieved by physical means. Summarising, a fixative should:

(1) preserve cells from deterioration;
(2) stabilise the cell's shape and structure;
(3) prevent losses of cell contents;
(4) make available reactive sites for linking with colouring agents (to this must be added: leave the charge on the reactive sites so that they can become dyed differentially by anionic and cationic dyes);
(5) make the cell membrane permeable for the dyes.

In cytology, the most common fixatives contain ethyl alcohol. To distinguish these methods from the other common cytological fixation method of air drying, they are called 'wet-fixation methods'.

4.2 Requirements of a Fixative

4.2.1 Cell Preservation

In all fixation methods preservation and fixation are both achieved at the same time. Therefore, many fixatives are also used as preservatives. The autolysing enzymes are inactivated by the fixatives. Some other enzymes which are used as histochemical agents in enzyme staining may be left intact. We are not dealing with this type of staining in this book so we shall not discuss the problems arising when some enzymes are to be inactivated and others not.

4.2.2 Stabilising Shape and Structure

Most fixation methods change the size and shape of cytoplasm and nucleus considerably, especially in cytological techniques (see chapter 2). However, if the fixation is performed in a consistent way, these changes, although they vary between cell types, vary little within one cell type. Maintaining the cell's structure can be achieved by 'freezing' the position of the cell components at the moment of taking the sample. This can be done by rendering them insoluble. The most important and most effective candidate for such treatment is protein, forming the main component of the cell. Protein can be made insoluble in two ways:

(1) Cross-links between the peptide chains are formed, leaving the secondary and tertiary structure intact. This results in a macromolecular network consisting of very large protein aggregates. In cytological preparations they precipitate on the glass slide.
(2) The protein is coagulated, with denaturation as its extreme form. The secondary and tertiary structures are disrupted by interference with the hydrophobic bond between the lipophilic sides of the peptide chain. Coagulant fixatives like the alcohols and acetone, and certain metal fixatives like mercuric chloride, have such effects. These can also be achieved by physical treatment such as heat or cold, or by air drying.

Figure 4.2 shows the effect of some fixatives on films of protein (gelatin/albumin). Alcohol clearly

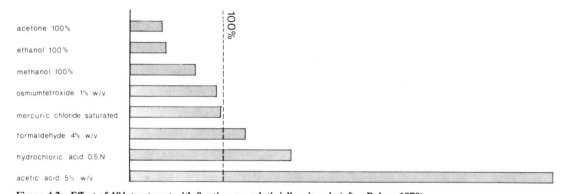

Figure 4.2   Effect of 18 h treatment with fixatives on gelatin/albumin gels (after Baker, 1970).

Table 4.1 Nuclear and cellular size and nuclear/cytoplasmic ratio of pyriform benign urothelial cells subjected to various cytopreparatory techniques

	Nuclear area (μm)	*Cellular area (μm)*	*N/C ratio*
Air dried, May-Grünwald–Giemsa	95 ± 11	277 ± 65	0.37 ± 0.07
50% Ethyl alcohol, Papanicolaou	49 ± 9	176 ± 46	0.30 ± 0.07
96% Ethyl alcohol, Papanicolaou	32 ± 6	114 ± 56	0.29 ± 0.06
Leiden spray fixative, Papanicolaou	40 ± 9	131 ± 42	0.31 ± 0.07
No fixation, no staining, phase contrast	60 ± 12	208 ± 72	0.34 ± 0.06
Methanol–acetic acid	78 ± 8	—	—

has a very strong shrinking effect whereas formaldehyde makes the protein swell. Acetic acid has by far the strongest swelling effect. It is added to many fixative mixtures for this reason.

The shrinking effect of the coagulant fixatives can be explained by the squeezing out of water molecules from the hydrophilic side of the peptide chains which are kept there in the tertiary structure. The altered structure after fixation is again due to the reaction of proteins. In addition, fixation, especially with alcohol, stiffens the cell. The fine macromolecular network caused by cross-linking agents, and the structure of denatured proteins caused by coagulant fixation, are both very different from the cell's original structural features.

Each fixation also changes the chromatin pattern in its own way. This is due to the difference in denaturation of the nucleoproteins (see section 4.2.4), a mechanism that plays an important role in the staining of DNA with nuclear dyes.

Fixation changes nuclear and cytoplasmic size: these changes differ from fixative to fixative (see table 4.1). An additional factor in size changes in cytopreparatory techniques is the spreading of the

nucleus and cytoplasm over the glass slide (see chapter 2). If the cell is somewhat stiffened by the fixative, the spreading is less than might be expected.

4.2.3 Prevention of Loss of Cell Contents

During fixation and throughout the staining process, cells are subject to loss of their contents. This happens especially during the hydrating and dehydrating stages. For the cell components that interest us most of all, the proteins and nucleic acids, some data are available. In general it can be said that the cross-linking fixatives are better at keeping the proteins within the cell than are the coagulating agents. The macromolecular network formed by cross-linking traps the soluble proteins, glycerol and lipids in its meshes. Less is known about the loss of nucleic acids. Acid fixatives retain DNA quite well within the nucleus, hence the use of Carnoy's fluid in studies of this organelle. This contains alcohol and acetic acid. RNA is also well retained by Carnoy's fluid but is easily washed out in subsequent washes during dehydration. An extreme form of loss of cell contents is lipid extraction by alcohol fixation.

4.2.4 Revealing Reactive Sites for Linking with Colouring Agents

4.2.4.1 Colouring Proteins

The disruption of their secondary and tertiary structure can expose those reactive sites in proteins which were obscured before fixation. Fixation should leave the balance of the positive and negative charges the same (or nearly so) so that the proteins become colourable by generally used dyes. In figure 1.9 we showed the titration curve of a typical protein. After fixation the basic proteins should remain basic and the acid ones acid, so that they can be dyed by anionic and cationic dyes, respectively. If one wants to emphasise the acid or basic character of a protein and make use of this for a special dyeing technique, one should alter the pH of the fixative accordingly (see chapter 8).

4.2.4.2 Colouring Chromatin

The structure of chromatin is such that the reactive groups of DNA (phosphate groups) are covered by the nucleoproteins (see section 1.1.1 and figure 1.1). A good fixative should split off DNA from the protein core, so that the phosphate groups can react with nuclear dyes. There are differences between the degree to which this process occurs in coagulant fixation with alcohol and in fixation by air drying. The staining by Haematoxylin of an air-dried nucleus is much less intensive than that of a corresponding alcohol-fixed one. All reactive sites are not yet revealed by air drying. In these air-dried nuclei, additional reactive sites are revealed by post-fixation with methanol. This is the reason why smears for the Romanowsky–Giemsa method *must* be post-fixed prior to staining. However, after the customary fixation time of 15 min in methanol, the process of revealing reactive sites of the DNA is not yet fully completed. Longer fixation times in methanol will therefore change the chromatin staining somewhat. Incubating air-dried slides overnight with saline formol (van der Griendt's

fluid) also causes more reactive sites of the DNA to be revealed. The smears can then even be stained by the Papanicolaou method with satisfying results (see Atlas, plate 8). The chromatin pattern is then not identical to, but resembles closely, the pattern of chromatin in directly wet-fixed cells.

4.2.5 Making the Cell Membrane Permeable for Dyes

Ranvier's categoric statement that the cell must be fixed before it can be dyed is based on the fact that the living cell keeps most dyes out. An exception is made by vital dyes, which owe their colouring action to the fact that they are first dissolved in the lipids of the membrane before they can enter the cytoplasm (see chapter 8). However, most dyes used in cytology do not act that way: they have to penetrate the cell directly. This is achieved by fixing the cell. Fixation destroys the cell membrane so that the dyes can enter the cytoplasm. In addition,

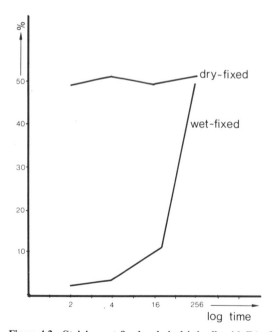

Figure 4.3 Staining wet-fixed and air-dried cells with EA of smears taken on the same day of the menstrual cycle. Over a longer period of time, a larger proportion of wet-fixed cells stain with Eosin (red). In the air-dry method, this is already achieved after a staining time of 2 min.

fixation can change the pores of the membrane in different ways (Horobin, 1982). For instance, air-dried cells are probably more 'cracked'. This can result in different staining patterns of air-dried and wet-fixed squamous cells using polychrome cytoplasmic dyes with customary staining times: the 'cracked' air-dried cells stain red in much higher percentages with Eosin in the Papanicolaou method. These differences disappear when extremely long staining times are applied (figure 4.3).

4.3 Rate of Penetration of Fixatives in Wet Fixation Methods Performed on Slides

The different fixatives do not penetrate into tissues at the same rate. Experiments with protein models and with liver tissue have shown the following order from slowest to fastest: ethanol, osmium tetroxide, mercuric chloride, formaldehyde, acetic acid. In cytology the rate of penetration is less critical. Ethyl alcohol, which ranks as a slowly penetrating agent, is the most widely used fixative in cytology. When alcohol is mixed with polyethylene glycol, as is the case in spray fixatives, the rate of penetration is slightly less, with a concomitant slight change in chromatin pattern: the distinction between hetero- and euchromatin is less. The higher the percentage of the alcohol solution, the faster its penetration, resulting in an increase of stainability of heterochromatin.

4.4 Fixation by Air Drying

4.4.1 Rate of Drying

In general, the Romanowsky–Giemsa staining method is applied on air-dried preparations post-fixed with methanol (see section 4.2.4.1). For optimal results, the cells should be dried very fast. This is the case when the cells are thinly spread over the slide, and when the material does not contain too much mucus. Only then does the 'Romanowsky effect' occur (see chapter 9). The drying process can be speeded up by using a hair-drier or putting the slides in a warm place. The final desiccation of the

cells should take place in a physiological environment.

Most sediments contain no albumin but a high content of electrolytes (as is the case in sediments from urine or cerebrospinal fluid). Then the electrolyte content of the evaporating overlying fluid will sharply increase in the air-drying process, and the final desiccation of the cell will happen in a highly hypertonic environment. The result is that the nuclei 'explode'. To prevent this from happening 20% albumin can be added to the sediment, or a cytocentrifuge or sedimentation chamber can be used. In both methods the fluid containing the electrolytes is slowly absorbed by filter paper while the cells settle down and dry. Also, when the cells are slightly pre-fixed with 80% methanol, during the drying process the chance that the nuclei will explode is less. Cells that dry in a physiological environment, such as blood serum, seldom explode. Methods to prevent explosion can be summarised as follows:

> Shorten the last step (use hair-drier, warm oven).
> Absorb supernatant in filter paper.
> Add 20% albumin.
> Add 80% methanol to supernatant.
> Remove mucoid material.

When cells are dried slightly prior to wet fixation, the air-drying effects are visible in the Papanicolaou stained smear. These cells have a decreased nuclear stainability and less 'crisp' chromatin pattern (see section 4.2.4). Smears containing little or no mucus and cells without protective squamous cytoplasm are especially prone to suffer from air-drying effects.

4.4.2 Post-fixation of Air-dried Smears

It is important to remember that, prior to staining, the air-dried slides should be post-fixed in an alcohol solution. Methanol (see chapter 9) is most frequently used for this purpose. Methanol post-fixation results in some difference in staining of heterochromatin and euchromatin, causing a different chromatin pattern (see also section 4.2.4.2) from that in directly wet-fixed cells in that deposition of chromatin beneath the nuclear envelope is absent and the contrast is less obvious. The longer air-dried cells remain in the fixative, the more the nuclear patterns will resemble those of directly wet-fixed ones. This is, for instance, visible in plate 9.4 of the Atlas: here the slides were in the methanol-containing staining bath for a period of 45 min, resulting in a more 'open' chromatin pattern.

4.5 Fixation of Cells in Suspension and Other Preservation Methods

Cells in suspension can be pre-fixed in order to prevent cell deterioration. The pre-fixed cells are finally fixed on the slide with another fixative, or with the same fixative as used for pre-fixing. In most methods, alcohols, with or without polyethylene glycol, are used. The disadvantages of pre-fixing with alcohol, especially ethyl alcohols, are: precipitation of proteins in the fluid; hardening of the cells, interfering with their capacity to attach to the slide (see chapter 2); and condensation of chromatin, making the nuclei susceptible to over-staining with Haematoxylin. In most methods, 50% ethyl alcohol is used; in Saccomano's method (see appendix 1) this is combined with polyethylene glycol. Some preservatives contain methanol instead of ethanol. Various methods are described in chapter 10.

A bacteriostatic agent is sometimes added (see figure 4.1).

4.6 Polyethylene Glycol in Wet Fixation Techniques

Still reported in 1953 (Hadju, 1983) that polyethylene glycol can be added to ethyl alcohol in fixatives. It can also be added to mixtures containing ethyl alcohol (Leiden fixative). The exact role of polyethylene glycol is not known. When used in spray fixatives, its main function is to form a film over the alcohol so that it cannot evaporate prematurely. Saccomano *et al.* (1963) reported the use of polyethylene glycol in fixing sputum in mixtures of it with 70% alcohol. In this method, the polyethylene glycol probably also enters the cells. Since Still was the first to use polyethylene glycol in

fixatives, and Saccomano made its use popular, Hadju (1983) proposed the use of the term 'Still–Saccomano technique'.

Most commercial fixatives contain polyethylene glycol. Saccomano used no. 1530, which is hard; we prefer to use no. 300, which is soft and quite easy to handle (see Leiden fixative, Appendix 1). Prior to staining, the polyethylene film must be removed from the slides. It is soluble in water.

4.7 Fixation Times

Wet fixation times can be too short (under 15 min), but never too long. When wet fixation times are too short, the desired changes in the nuclear proteins will not have developed, and thus the desired chromatin patterns are not visible. The cytoplasm will appear ill-defined and the nuclei will be stained too pale. If a fast evaluation of the cell sample is desired, a compromise is made between speed and quality (see chapter 8).

4.8 Fixatives: their Reaction with Cell Components; their Effect on Some Cell Features and on Staining Results

For histological techniques an impressive list of fixative mixtures is available, each best suited to a specific purpose. Of these only a few have found their way into routine cytopreparatory techniques:

(1) the non-coagulant cross-linking agents: osmium tetroxide, formaldehyde, glutaraldehyde and acetic acid (usually in combination with other fixatives);
(2) the coagulant fixatives: ethyl alcohol in various grades, methanol–acetone (usually as an addition to the alcohols) and mercuric chloride (in saturated solution).

Apart from the distinction between coagulant and cross-linking, one can distinguish between additive and non-additive fixatives. Additive fixatives link themselves with the substrate to be fixed. They can be either coagulant or non-coagulant. An additive fixative that does coagulate is mercuric chloride, and to the additive fixatives that do not coagulate

Table 4.2 Primary fixatives and their effects on cell components

Primary fixative	Effect on: Proteins	Nucleoproteins	Nucleic acids	Lipids	Glycogen
Ethanol, C_2H_5OH non-additive coagulant	Denatures most; gelatinises histones; charge not altered or slightly altered	Does not coagulate	Denatures; solvation of bases increased; solvation of riboses decreased	Dilute ethanol splits off lipids from lipoprotein complexes	Precipitates but leaves it unchanged
Mercuric chloride, $HgCl_2$ additive coagulant	Cross-links; reacts with ionisable bases, with SH group	Coagulates into fine flocculi	Cross-links through reaction with N of heterocycles; precipitates DNA weakly; nuclear mercuration possible	Does not insolubilise; hydrolyses plasmalogen; useful before applying Schiff's reagent	Not known
Formaldehyde, $\begin{smallmatrix} \diagup OH \\ C=O \\ \diagdown OH \end{smallmatrix}$ additive coagulant	Cross-links through reaction with NH_2 and SH groups forming methylene bridges	Do not coagulate	—	Preserve well	Does not change glycogen but traps it in macromolecular protein network
Glutaraldehyde, additive coagulant	Renders charge less negative; gelatinises histones				
Osmium tetroxide OsO_4 additive non-coagulant	Cross-links through reaction with NH_2 and SH groups; renders charge less positive	Does not coagulate	Can react with $C=C$ groups of bases giving diol derivatives; does not precipitate DNA	Reacts with double bonds, giving black colour	Little or no reaction

belong the aldehydes and osmium tetroxide.

In table 4.2 we have listed the primary fixatives that we have mentioned in our discussion and their reaction with proteins, nucleoproteins, nucleic acids, lipids and glycogen.

The various methods of fixation each have their effect on the stainability of the cell components and other features of the cell. The difference in staining effect is summarised in figure 4.4, where the results are shown of liver cells fixed by several well known fixative mixtures and stained by the two best known cytological stains, Papanicolaou and May-Grünwald–Giemsa. The effect of fixatives on cell features such as the contrast between euchromatin and heterochromatin, the visibility of cell and nuclear membrane and the distribution of chromatin in the nucleus is illustrated in the figure. It is obvious that some fixative and staining-method combinations are superior to others, as will be seen from figure 4.4

In Appendix 1 we give the recipes for those fixatives and fixative mixtures which are used in cytology. The choice of fixative is determined by a number of factors:

(1) Which staining method is going to be used? For instance, Romanowsky–Giemsa staining is disastrous after fixing with formalin (the aqueous solution of formaldehyde). Even the slightest trace of formaldehyde vapour in the laboratory spoils the staining. Sex chromatin cannot be visualised in air-dried cells. Papanicolaou staining is optimal after wet fixation.

(2) What is the purpose of the cytopreparatory procedure? Osmium tetroxide leaves the cell very much in its original state, which makes it a

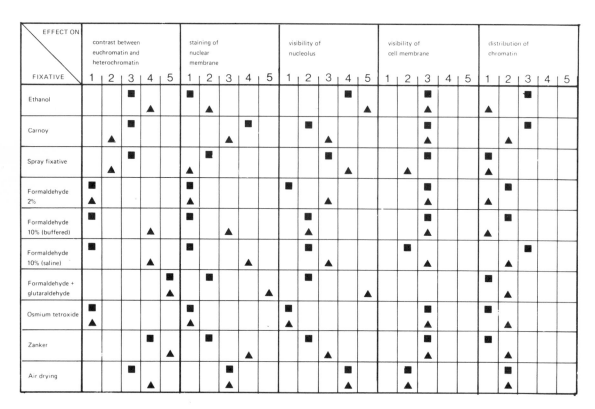

Figure 4.4 The effect of a number of fixatives on some cell features using two different stains. The composition of the fixative mixtures can be found in Appendix 1. The cells are obtained from liver imprints. ▲ = Papanicolaou, ■ = May-Grünwald–Giemsa. 1–5 grading as follows: 1, little–much; 2, none–well stained; 3, none–good; 4, clear–unclear; 5, even–uneven.

very suitable fixative for electron microscopy, but it does not display the difference between hetero- and euchromatin, and therefore does not show minute changes in chromatin pattern important in cytodiagnosis. Mercuric chloride is a good fixative of chromatin so it is popular in chromosome studies.

(3) Convenience. Fixation with ethyl alcohol with or without polyethylene glycol is a very simple way of preparing cells for a number of techniques. Mercuric chloride requires after-treatment with iodine and thiosulphate, which makes it a complicated method.

Whatever fixative one chooses, the result always shows 'artifacts', but as long as they are wanted and constant, it does not influence cytodiagnosis in a negative way and may even enhance it. As has been said before, osmium tetroxide leaves the cell very much in its original state but this cannot be said for any of the other fixatives. To achieve a consistent result the cells should be fixed in a consistent way, and when air drying is used, the cells should be in a *physiological environment*. The air-drying method and the wet-fixation method should *not* be applied in combination, as can happen accidentally when the alcohol reaches the cells when they are partly dried ('too slowly fixed').

In this discussion on fixation we have not gone into all the technical details which are known from histochemistry. For further reading we recommend Baker, Hopwood, Pearse and Horobin for their accounts on the various aspects of fixation and fixatives (see the list of recommended reading at the end of the book).

5

Nuclear Dyes

5.1 General Properties

From chapter 1 it has become clear that the nucleus contains acid and basic components which are separated from each other by some kind of fixation. Thus both acid and basic components can become available to react with dyes. In section 3.7 we have seen that, through ionic bonding, acid tissue components can link with basic dyes and basic components with acid dyes.

Ethanol (in the index referred to as ethyl alcohol), which is generally used as a fixative in cytology, has the advantage of changing the isoelectric point of proteins little if at all. This means that, at the pH which is normally used, the nucleic acids and the non-histone proteins react as acids and the histone proteins as bases, yet acid dyes are not often used as nuclear dyes although they do occur as such in some staining methods. In Mallory's tricolour method the Acid Fuchsin dyes the nucleus, and in Heidenhain's Azan the Azocarmine does the same (Lillie, 1977).

Since, in routine cytology, the main objective of staining cells is to provide a clear differentiation between nucleus and cytoplasm, basic dyes are used to colour the nucleus, leaving the cytoplasm, which is predominantly basic in reaction, colourable by acid dyes. Although the basic components of the nucleus will also take on the acid dyes, this is masked by the stronger colouring caused by the basic dyes.

In the following sections we shall divide the nuclear dyes into general and specific. The general dyes serve to distinguish between nucleus and cytoplasm, and the specific dyes stain DNA and/or RNA, or the nucleoproteins. Dyes used to colour the nucleolus are discussed in chapter 7.

5.2 General Nuclear Dyes: Haematoxylin

Because of its long history as a nuclear dye and its widespread use, we shall discuss Haematoxylin first and in great detail. Haematoxylin is the most frequently used natural dye in microtechnique. Because of this, and also because its manufacture and use offer such a good insight into the principles of staining, it deserves special attention.

5.2.1 The Active Component of a Haematoxylin Solution

Haematoxylin is obtained from the American Logwood (*Haematoxylon campechianum*). From its chemical formula it can be seen that it cannot be a dye in the sense described in chapter 3, because it does not possess any of the necessary chromophores. When dissolved in water, it has a light yellowish colour, but if left for several weeks, the colour of the solution gradually changes to red and then dark red. This process is extensively described by Heidenhain (1896) and is attributed to contact with the air above the solution, to the effect of light, and most of all to contact with the glass of the container. Mayer (1891) had already explained the time of 'ripening' (as the change is called) as a process of oxidation by atmospheric oxygen. It is the oxidation that provides the Haematoxylin with the necessary chromophore (of the quinoid type). The oxidation product of Haematoxylin is a Haematein. This is the actual dye, and not Haematoxylin as such. Thus, properly speaking, the nuclear dye is not Haematoxylin but Haematein. However, it is usual to apply the name Haematoxylin to the dye and we shall do the same.

Natural oxidation, by the oxygen in the air above the solution or trapped in the solution itself, is a slow process. It can be hastened by the addition of oxidising agents. A number of these have been compared by Clark (1974) and they all give similar and equally good results. Sodium iodate is most commonly used, but Harris (1898; see Bolles-Lee, 1900) prefers mercuric oxide. A very quick way of ripening is leading pure oxygen through a Haematoxylin solution, as advocated by Hance and Green (1961).

It would seem logical to use Haematein from the onset instead of starting off with Haematoxylin. The same idea occurred to Heidenhain (1896) but his results with Haematein were not superior to those with Haematoxylin. Moreover, the Haematein solution had a much shorter lifespan. This can be explained by the continuing oxidation process which does not stop at Haematein but carries on to form oxyhaematein and further. The oxidation products tend to flocculate and make the solution unsuitable for further use.

The amount of oxidising agent determines the way in which the dye is being used. According to Baker (1970), 0.187 g $NaIO_3$ is needed to oxidise 1 g of Haematoxylin. This ratio is adopted by Gill *et al.* (1974), who produced half-oxidised Haematoxylin by using half the amount of oxidant. Harris's Haematoxylin contains an amount of oxidant that oxidises about a third of the Haematoxylin, leaving the remainder to oxidise naturally. Mayer uses the oxidant–Haematoxylin ratio as mentioned by Baker (1970) for complete oxidation.

5.2.2 The Mordant in the Staining Solution

Although the staining solution obtained through oxidation of Haematoxylin has a proper dye in it, it is still not suitable to stain the nuclei for which it is intended. Because of its DNA and RNA content,

the nucleus has anionic groups available for reaction. The positive charge of the amphoteric Haematein is not strong enough to form a direct bond with the negatively charged groups of the nucleic acids. In order to link dye and tissue, a *mordant* is needed. From its formula it can be seen that the Haematein molecule has a suitable place to form a chelate with a metal atom, as explained in section 3.9.2. Aluminium alum is most commonly used for this purpose. Iron and chromium are less often applied. One's preference will largely depend on the colour one wants to obtain. Aluminium gives a blue result whereas iron and chromium produce black or dark-grey nuclei, respectively.

5.2.3 Preparing the Haematoxylin Solution

Table 5.1 lists the various ways in which Haematoxylin solutions can be made. No doubt there are many more variations on the same theme. The choice may be one of convenience, taste or number of slides to be stained. It is worth experimenting to find one's own preferred formula.

Points for consideration when preparing a Haematoxylin solution are well summarised by Cole (1943) and Gill *et al.* (1974). Heidenhain's description of his experience in the laboratory also makes very good reading if only to stimulate further experimentation in the art of dyeing (Heidenhain, 1896). Some Haematoxylins stain the nucleus differently from others (table 5.1). There are many commercial Haematoxylins on the market; however, they are far more expensive than the home-made ones (for recipes see appendix 3). Additives may be added to a Haematoxylin solution (Clark, 1975). Addition of acid, glycerine and ethyl alcohol makes the solution more stable, and the alcohol also serves to prevent mould growth. The 'brake' effect of acid can be reversed by the addition of alkali.

5.2.4 Progressive versus Regressive Staining

In *regressive staining* with Haematoxylin excess dye is removed by acid (see section 3.10.2). In addition, dye which has settled in the cytoplasm, colouring it blue, is removed. Since Haematein becomes red at

Table 5.1 The history of Haematoxylin

Author	Haematoxylin per 1000 ml (g)	Oxidant	Mordant	Stabiliser (ml)	Solvent (ml)	Remarks
Delafield, 1885 (Bolles-Lee, 1900)	3.2	spontaneous	Al-alum saturated (1:11) 320 ml	96% alcohol, 20 methanol, 80 glycerine, 80	dist. water, 500	very stable; used regressively
Ehrlich, 1896 (Bolles-Lee, 1900)	6.5	spontaneous	Al-alum in excess	100% alcohol, 320 glycerine, 320 glacial acetic acid, 32	water, 320 100% alcohol, 320	very stable; used regressively
Mayer, 1891 (Bolles-Lee, 1900)	Haematein, 1.0	NaIO$_3$, 0.2 g	Al-alum, 50 g	100% alcohol, 50	dist. water, 1000	used progressively
Heidenhain, 1892 (original)	5	spontaneous	Fe-alum 2.5%	100% alcohol, 100	dist. water, 900	mordant solution used for differentiation; used regressively
Unna, 1892 (Bolles-Lee, 1900)	3.3	spontaneous	Al-alum, 33 g	100% alcohol, 330 sublimed S, 2 g	water, 660	used regressively
Hansen, 1895 (Romeis, 1968)	4.7	KMnO$_4$, 0.88 g	K-alum, 93.9 g	100% alcohol, 47	dist. water, 940	KMnO$_4$ serves as disinfectant; used regressively
Harris, 1898 (Romeis, 1968)	4.7	HgO, 2.4 g	Al-alum 95.0 g	100% alcohol, 48	dist. water, 952	used regressively
Mallory, 1897 (Bolles-Lee, 1900)	1.0	spontaneous	phosphotungstic acid, 10 g	—	water, 1000	used progressively
Mallory, 1901 (Bolles-Lee, 1937)	1.0	H$_2$O$_2$, 2 ml	phosphotungstic acid, 20 g	—	water, 1000	used progressively
Mayer, 1903 (Bolles-Lee, 1937)	1.0	NaIO$_3$, 0.2 g	Al-alum, 500 g	—	dist. water, 1000	used progressively
Weigert, 1904 (Romeis, 1968)	5.0	FeCl$_3$ 5.8 g	FeCl$_3$	96% alcohol, 500 HCl, 5	dist. water, 490	used regressively

Table 5.1—*continued*

Author / Ingredients per 1000 ml	Haematoxylin (g)	Oxidant	Mordant	Stabiliser (ml)	Solvent (ml)	Remarks
Unna, 1910 (Bolles-Lee, 1937)	5.0	H_2O_2, 100 ml	Al-alum, 33.0	100% alcohol, 100 glycerine, 200	water, 600	used regressively
Cole, 1943 (original)	0.5	iodine, 0.5 g	Al-alum, 70 g	96% alcohol, 50	dist. water, 950 ml	used progressively
Gill *et al.*, 1974 (original)	2.0	$NaIO_3$ 0.2 g	Al-sulphate, 17.6 g	glacial acetic acid, 20 ethylene glycol, 250		Haematoxylin half-oxidised by calculated amount of Na-iodate

very low pH values, the nuclei remain red when the slide is not treated after the acid bath. This unwanted effect is counterbalanced by placing the slide in a alkaline solution. This is called '*blueing*'. Ordinary tap water can be used for this purpose if its pH is slightly higher than 7. Otherwise a saturated solution of lithium carbonate can be used, or Scott's solution. In *progressive* staining, the slides receive the correct amount of Haematein and thus differentiation is unnecessary but proper rinsing is mandatory. Needless to say, progressive staining solutions contain less Haematein than those that stain regressively.

5.2.5 What is Stained by Haematoxylin?

Haematoxylin stains both DNA and RNA and, in addition, acid nucleoproteins. This is in contrast to some other nuclear stains like, for instance, Feulgen's reagent which only stains DNA. If excess Haematoxylin is deposited in the cell, it can be trapped in the cytoplasm. This is unwanted staining and should be removed by means of an acid bath, either aqueous or alcoholic, and thorough rinsing. The nuclei are always stained a darker blue than the cytoplasm so that they can be recognised quite easily even when staining with Haematoxylin only ('monostaining').

5.2.6 Additional Nuclear Staining by Cytoplasmic Stains in Haematoxylin-stained Slides

Anionic dyes, which are mainly used as cytoplasmic dyes in routine cytological techniques for light microscopy, colour the basic proteins. These are also present in the nuclei (the histones in chromatin, see section 1.1.1), so these dyes also stain the nucleus to some extent (see also section 6.3.2.2). Condensed nuclei are especially likely to take up the cytoplasmic dyes. This is the reason why, in Papanicolaou-stained smears, pyknotic nuclei are more red (by staining with Eosin Y) than vesicular nuclei.

5.3 Other General Nuclear Dyes

Being a natural dye the supply of Haematoxylin is not always sufficient and there have been times when there was an actual shortage of it. Such times have been very useful in the past, especially in the early 1970s, in stimulating research into the possibilities of other nuclear dyes. Also the introduction of automation of staining and diagnostic methods has increased the need for other nuclear dyes which are more stable than Haematoxylin. Quantitative cytochemistry also requires stable dyes producing constant reproducible stoichiometric staining (see section 5.4.1). The majority of general nuclear dyes are used with a mordant. This is not because they would otherwise not stain the nucleus but because the bond dye–mordant–substrate is a stronger one than the direct bond. We shall mention a number of Haematoxylin substitutes here which have been tested on histological material. However, this is no guarantee that they are also suitable for cytological samples.

Solochrome Cyanin R (Pearse, 1957; Llewellyn, 1974; Gurr, 1975; Clark, 1980) has the pleasing feature of being able to stain cytoplasm and nucleus differentially, but unfortunately its use for cytological purposes has not been reported to be successful yet. We have tested the dye for staining gynaecological samples directly and found the nuclei rather delicately stained with a fairly strong colouring of the cytoplasm.

Rhodanile Blue and Pontacyl Black SX (Amido Black 10B) were also reported by Gurr (1975). In our tests Rhodanile Blue showed a pleasant blue colour for cytoplasm and nucleus, the latter being darker but showing little detail. Pontacyl Black SX can also be used as a typical stain for nucleoli (see chapter 7).

Lillie *et al.* (1975a, 1976a) tested a large number of nuclear dyes of which some were successful and others not. Phenocyanin TC, Gallein, Alizarin Cyanin BB, Coelestine Blue, Carmine and Gallamin Blue S were the most successful dyes. Gallocyanin, which is also used as a typical RNA dye (Brown and Scholtz, 1979), has the disadvantage of needing a very long staining time (24 h).

Other Haematoxylin substitutes worth mentioning are Iron Alizarin Blue (Meloan and Puchtler, 1974) and Iron Coelestine Blue (Gray *et al.*, 1956). The latter was found to be very satisfactory for cytological samples by Yasumatsu (1977).

Lillie *et al.* (1976b) add iron and aluminium lakes of Gallo Blue E to their list of nuclear dyes. This dye stains mucin metachromatically red at the same time.

Of the two related dyes, Fluorone Black and Methyl Fluorone Black, the former gave the best results as a nuclear dye. As iron mordant lake it stains nucleoproteins metachromatically (Lillie *et al.*, 1975b).

Of the thiazines, Methylene Blue, Toluidine Blue and Thionin are perhaps the best known as nuclear dyes. The latter two have been reintroduced as nuclear dyes in automated methods (D.H. Wittekind, personal communication). However, none of the thiazines shows as much nuclear detail as Haematoxylin or some of its substitutes mentioned above. It is generally thought that basic dyes like the thiazines link to the acid nuclear substrate by ionic bonding, whereas the metal complex dyes are bound to acid groups by covalent linking through the metal ion. This generalisation is doubted by Marshall and Horobin (1973a), who consider that at least some of the well known metal dye complexes are as large as basic dyes. This means that, for them, van der Waals forces are as important as hydrophobic forces in the linking mechanism.

5.4 Specific Nuclear Dyes

These can be divided into dyes for staining DNA, those that stain RNA and those that demonstrate nucleoproteins. In all cases the dyes are applied to allow quantitative assessment of the nuclear components.

5.4.1 Demonstration of DNA

Undoubtedly the best known method for specifically demonstrating DNA is the Feulgen reaction (see section 3.5.2.4). Quantitative assessment of the amount of DNA by cytophotometric measurements after Feulgen's reaction is not, however, absolute. The reaction of Schiff's reagent with the aldehydes of DNA is a stoichiometric one. There can never be a full guarantee that all possible binding sites have been made available by fixation although one should hope so. Notwithstanding these drawbacks, the Feulgen method is widely used as a quantitative stain.

Fluorochromes used in fluorimetry are also fairly reliable informants on the amount of DNA in the nucleus (Caspersson *et al.*, 1983). We have not included these in our survey and refer to Böhm and Sandritter (1975) for more information on this aspect of DNA measurement.

5.4.2 Demonstration of RNA

In the combination Methyl Green–Pyronin Y, the latter is supposed to dye RNA specifically. However, the dye is not specific for RNA but also stains the lower polymers of DNA. Kurnick (1952) considers it merely to be a qualitative counterstain for Methyl Green.

Quantitative staining of RNA by Gallocyanin is suggested by Brown and Scholtz (1979). They stained sections of RNA–gelatin blocks with Gallocyanin chrome alum at pH 1.64 and found a direct correlation between amount of dye and amount of RNA in the gelatin section. The stoichiometric relationship between Gallocyanin and RNA was examined earlier by Sandritter *et al.* (1954, 1966).

Specific RNA staining by thiazine dyes makes use of their metachromatic qualities by which RNA can, under specified circumstances, become stained metachromatically and chromatin orthochromatically. The success of such methods depends very much on the type of fixation, pH, ion concentration other than that of the dye (i.e. added salts), dye concentration and embedding medium. To overcome some of these problems, Bennion *et al.* (1975) suggest the use of basic dyes in combination with surfactants. In their experiments the most successful dyes were the thiazine dyes Azure A and Toluidine Blue. At pH values between 4 and 7, RNA-rich sites stained strongly purple (metachromatic) whereas other cell components, including chromatin, stained pale blue. In the combination dye–surfactant, the latter is present in great molecular excess and its molecules are aggregated

into micelles. The staining result is supposed to be the result of competition between dye ions which are partly aggregated and the surfactant micelles. Those cell components which are most difficult to penetrate (i.e. RNA) are stained by the dye ions and dye aggregates which are still smaller than the surfactant micelles. Apparently, enough dye ions can reach these sites to cause metachromatic staining. With this method other cell components are stained pale blue. It appears that the dye concentration at those sites is not high enough to cause metachromatic staining because of dilution with surfactant micelles.

5.4.3 Demonstration of DNA and RNA

To differentiate between the two nucleic acids, pairs of basic dyes can be used of which one links to DNA and the other to RNA. To this end the combinations Methyl Green–Pyronin Y, Acridine Red–Malachite Green and Thionin–Pyronin have been successfully used.

The mechanism of the preferential uptake of the basic dyes by the nucleic acids is not always known. In the case of the better known combination, Methyl Green–Pyronin, the dyes appear to distinguish between the different states of polymerisation of the nucleic acids (Kurnick, 1952). The method is very old as a differential stain (Unna, 1902). It has obtained a new lease of life as a histochemical stain. Whether it can serve as a quantitative method for DNA and RNA content of the nucleus depends on the method which is used.

The most common method for using the dye combination is to stain the cells in a mixture of Methyl Green and Pyronin in water, followed by a rinse in water, dehydration in acetone and clearing in cedar oil and xylene. However, the rinse in water extracts the Methyl Green, and acetone removes Pyronin, so that the method is unsuitable for quantitative assessment. Taft (1951) improved the method by using a buffered aqueous mixture followed by rinsing in water and differentiation in a mixture of tertiary butanol and absolute alcohol. This solves the problem of Pyronin removal but still allows Methyl Green to be extracted by water.

Kurnick (1952) suggests a modified version of Taft's method by staining the cells first with Methyl Green in acetate buffer (pH 4.2) or water, rinsing in *n*-butyl alcohol and then staining with Pyronin in acetone. After staining with Pyronin he does not rinse the cells but transfers them directly to cedar oil and then to xylene. The method leaves cytoplasm red or pink, nucleoli pink (depending on the Pyronin concentration), chromatin bright green and erythrocytes brown. The method can serve for quantitation of DNA but only as a qualitative method for the demonstration of RNA.

The sequence Schiff's reagent/Methylene Blue is reported by Spicer (1961) to provide a good differentiation between DNA and RNA after suitable fixation. To obtain the desired effect the cells need to be fixed in Bouin's fixative, which contains saturated picric acid, formalin and acetic acid. The fixative plays a double role: first, it fixes the cells, and secondly it hydrolyses DNA to enable it to react with Schiff's reagent as in the Feulgen reaction. Subsequent dyeing with a thiazine dye like Methylene Blue at a pH between 3 and 4 specifically stains RNA. The result is red DNA and blue nuclear and cytoplasmic RNA. The low pH of fixative and dye solution ensures typical RNA staining, leaving the acid nuclear proteins unstained.

An interesting way of specifically staining RNA is described by Mendelson *et al.* (1983). Cuprolinic Blue without the addition of cation stains both DNA and RNA. But when cations in the form of mercuric chloride are added to the dye solution, only the single-stranded RNA is dyed stoichiometrically (see also section 7.2.1).

5.4.4 Demonstration of Nucleoproteins

Just as important as the nuclear DNA content may be the protein content of the nucleus. When a cell is activated from its resting phase (G_0) to the first stage of activity (F_1), its protein content increases before there is any change in the amount of DNA. It is not until later stages (S-phase) that DNA synthesis is increased.

To measure both DNA and protein content of cells, Caspersson *et al.* (1983) use a Feulgen/Naphthol Yellow S sequence originally introduced into cytochemistry by Deitch (1955) and later developed by Gaub *et al.* (1975). Gaub's experiments were in the first place aimed at evaluating the relationship between DNA and nuclear protein. However, Naphthol Yellow S binds to cytoplasmic as well as nuclear basic proteins.

By using scanning cytophotometry it is possible to determine the amount of dye bound in cytoplasm and nucleus. Such measurements show the relation between Naphthol Yellow S and the dry mass of the cell, which can be used as a parameter for cytoplasmic and nuclear basic protein provided fixation and staining conditions are strictly controlled. Tas *et al.* (1980) tested a number of acid dyes on their qualities as protein dyes especially for cytophotometry. Their survey included Naphthol Yellow S, Fast Green at alkaline pH, Coomassie Brilliant Blue and Dinitrofluorobenzene. Such methods can only demonstrate certain reactive groups of the amino acids and cannot identify proteins as a whole. At the most, it is possible to find a ratio of dye to reactive group that may be representative for a certain protein or part of it. The experiments by Tas *et al.* were not particularly aimed at nucleoproteins. However, by using scanning cytophotometry, the absorbance in the different cell organelles can be measured and thus the mass of basic nucleoproteins can be established.

To differentiate between acid and basic nucleoproteins, one can use dye solutions at different pH. We have summarised some methods for staining nuclear components in table 5.2. There are of course many more variations on the same theme. The work of Smetana and Busch (1966) on the staining of acid proteins by Toluidine Blue at basic pH (9.0) is well known. Also Dhar and Shah (1982) report on pH-dependent staining of nucleoproteins. They were able to stain basic nucleoproteins by Fast Green FCF at pH 8.0 and acid nucleoproteins at pH 5.0. This is in agreement with the earlier findings of Alfert and Geschwind (1952).

In this review we have tried to give some insight into the problems of staining nuclear components without trying to present a complete picture of all the possibilities that exist.

Table 5.2 Some staining methods for nuclear components in relation to pH of the dye solutions

	1	2	3	4	5	6	7	8	9	10
Toluidine Blue (Smetana and Busch, 1966)									non-histone proteins from 9.0	→
Toluidine Blue or Azure A + surfactant (Bennion *et al.*, 1975)				RNA metachromatically						
Methylene Blue (Spicer, 1961)			RNA after Bouin fixation and Feulgen							
Gallocyanin Chrome Alum (Brown and Scholtz, 1979)	RNA									
Biebrich Scarlet (Spicer, 1961)								basic proteins from 8.0	→	
Fast Green (Alfert and Geschwind, 1952; Tas *et al.*, 1980; Dhar and Shah, 1982)					acid nuclear proteins pH 5.0			basic nuclear proteins pH 8.0		
Naphthol Yellow (Tas *et al.*, 1980)		total protein content pH 2.8								
Coomassie Brilliant Blue		total protein content pH 2.2								

6

Cytoplasmic Dyes and Stains

6.1 Introduction

Since the ground protoplasm of most cells is amphoteric in reaction, it can in principle react with both acid and basic dyes. It is possible to shift the existing balance of charges in the cytoplasm-proteins by changing the pH of the dye solution. Several staining methods are based on this possibility, especially those where quantitative measurement of protein is required.

A look back to table 3.1 shows that virtually all cytoplasmic dyes are acid, with the exception of Bismarck Brown. This dye formed part of Papanicolaou's original stain (Papanicolaou, 1942). Spectrophotometric measurements by Marshall et al. (1979), however, showed that the dye plays no part in cytoplasmic staining at all, which justifies its omission.

Cytoplasmic dyes are nearly always used as direct dyes. However, Puchtler and Isler (1958) claim that there is some form of mordant staining of Light Green in Papanicolaou's stain involving phosphomolybdic or phosphotungstic acid (see section 6.3.2.1).

In this chapter, the polychrome staining methods, including Papanicolaou's, are discussed in great detail. In addition, a selection of histochemical dyes and stains used in routine cytology to visualise components in the cytoplasm are given. The Romanowsky–Giemsa stains, colouring both cytoplasm and nucleus, are discussed in chapter 9.

6.2 Cytoplasmic Dyes

In the field of cytoplasmic dyes, Eosin Y takes up a similar position as Haematoxylin does among the nuclear dyes. It was originally used as a counterstain after Haematoxylin (Delafield, 1898—see Bolles-Lee, 1900). That combination is still the most widely used in histology. The only dye that can compete with Eosin in this respect is perhaps Biebrich Scarlet, although this dye is hardly ever used as a single counterstain but rather in combination with others like Fast Green (Shorr, 1942). Light Green and Fast Green are other familiar cytoplasmic dyes. As a single cytoplasmic dye Biebrich Scarlet is used by Spicer and Lillie (1961) for the identification of basic proteins. By staining with dye solutions of different pH they were able to differentiate between basic proteins according to their degree of basicity.

Much emphasis has been put on research into quantitative dyeing of proteins in recent years. Tas et al. (1980) report on a number of acid dyes which combine stoichiometrically with basic amino acid residues. Naphthol Yellow S, Fast Green FCF, Dinitrofluorobenzene and Coomassie Brilliant Blue were all found to be suitable dyes for measuring the amount of basic proteins in the cell. Oud et al. (1984) use the well known Light Green Y and Orange G for quantitative assessment of basic protein content, and emphasise the importance of the pH of the dye solution for maximum intensity of staining (see also sections 6.3.2.4 and 6.3.2.5).

The RNA content of the cytoplasm can also be demonstrated by using single dyes. Bennion *et al.* (1975) stained RNA-rich sites in the nucleus and cytoplasm metachromatically with the combination of a thiazine dye (Azure A or Toluidine Blue) and a surfactant. Spicer (1961) stained RNA differentially after Bouin fixation and Feulgen staining for DNA. RNA in the cytoplasm and nucleus became stained blue by subsequent staining with Methylene Blue at pH 3–4. Brown and Scholtz (1979) dyed RNA specifically with Gallocyanin Chrome Alum at very low pH (1.64).

In principle, if one is not interested in cytoplasmic differentiation as is obtained by Papanicolaou's method and by Shorr's, then a combination of a nuclear dye like Haematoxylin with either Eosin, Light Green or Biebrich Scarlet is sufficient in diagnostic cytology.

6.3 Polychrome Dyes

Polychrome dyes contain more than one dye, used at the same time. Polychrome staining was well known in histological techniques at the time when interest in its application in cytology first arose. Table 6.1 shows how, after various modifications, the single dye method has evolved from today's most widely used polychrome stain, Papanicolaou's EA (Eosin–Azure). In their original papers, Mallory (1900) and Masson (1929) described their polychrome staining methods for connective tissue which resulted in differential staining of the various components of the sections.

Papanicolaou's method as well as Shorr's are derived from these methods and have a differentiating staining effect on cells. Papanicolaou's cytoplasm stain comprises the dyes Eosin Y, Light Green Y, Orange G and Bismarck Brown. His method includes Haematoxylin as a nuclear dye. At about the same time that Papanicolaou developed this polychrome staining technique (1942), Shorr experimented with various cytoplasmic dye combinations to improve the existing methods of detecting different stages of cornification in squamous cells (Shorr, 1940*a, b*). He used a

modification of Masson's trichrome stain. The counterstain contains Biebrich Scarlet, Orange G and Fast Green. It results in contrasting colours in cervical smears.

Shorr's original method does not involve nuclear staining and can only be used for hormonal assessment of cells. See section 6.3.3 for a description of Shorr's polychrome stain.

6.3.1 Papanicolaou's Polychrome Stain: Origin and Development

The effect of ovarian hormones with their fluctuating blood levels on mammalian epithelium has been extensively studied since the early work of Papanicolaou. His original work, studying the menstrual cycle, concerned cells taken from the vagina of guinea pigs and later of women (1928, 1933). For the visualisation of the different stages of maturation of these vaginal cells, he devised a special method of fixation and staining. During his studies of vaginal smears from women, in one case he even encountered malignant cells. This led to a preliminary communication in 1928 at the Third Race Betterment Conference at Battle Creek, Michigan, which unfortunately was not noticed by the medical world of that time. In 1943 the monograph *Diagnosis of Uterine Cancer by the Vaginal Smear* (Papanicolaou and Traut, 1943) was published, but Papanicolaou's major interest remained focused on the study of the cyclic changes of the vaginal squamous epithelium.

In Papanicolaou's method, intermediate squamous cells stained blue. Towards the fourteenth day of the menstrual cycle the smears contained many superficial squamous cells. The cytoplasm of the latter stained red. The fourteenth day showed a peak of the proportion of red cells, which coincided with the peak in oestrogen activity. Thus it was possible to evaluate changes in hormone level by staining vaginal smears with a combination of dyes.

Although the combination of Haematoxylin, Eosin and Water Blue was satisfactory for hormonal assessment, it became clear that more cytological detail could be obtained by staining

Table 6.1 Development from connective tissue stain to Papanicolaou's EA

| | Connective tissue cytoplasm stains (g/100 ml) | | | | | | | | | | |
	Eosin	Acid Fuchsin	PMA	PTA	Oxalic acid	Aniline Blue/Water Blue	Orange G	Ponceau de Xylidine	Fast Green or Light Green Y	Biebrich Scarlet	Number of staining baths
Delafield (1889) (stain technology)	0.2–0.5		1								1
Mallory (1900) (original)		0.05–0.1	1		2.0	0.5					3
Masson (1929) ex Romeis (1968)		1.0 + glacial acetic acid 1 ml	1			Saturated + 2.5 ml glacial acetic acid	2.0				3
Papanicolaou (1933) original	0.5					0.5					2
Goldner (1938) (original)		0.03		3–5 g			2	0.06	Light Green, 0.1–0.2 in 0.2% acetic acid		3
Masson (1929) ex Goldner (1938)		0.3						0.6	Light Green, 1–2 in 0.2% acetic acid		3
Papanicolaou (1941) (original)	0.22	0.09	0.50			0.06	0.09				1
Papanicolaou (1941) (original)	0.2	0.1	0.47				1.2		Light Green, 0.6		1
Papanicolaou (1941) (original)	0.2	0.1	0.22	0.11		0.06	0.13		Light Green, 0.6		1
Foot (1938) (original)		0.03		3–5			0.06	0.06	Light Green, 0.1 in 0.2% acetic acid		3

Table 6.1—*continued*

Connective tissue cytoplasm stains (g/100 ml)

	Eosin	Acid Fuchsin	PMA	PTA	Oxalic acid	Aniline Blue/Water Blue	Orange G	Ponceau de Xylidine	Fast Green or Light Green Y	Biebrich Scarlet	Number of staining baths
Shorr (1940a,b) (original)		0.03		3–5			0.06	0.06	Light Green, 0.1 in 0.2% acetic acid		3
Shorr (1940a,b) (original)			2.5	2.5			0.4 in 1% acetic acid		Fast Green, 0.25 in 0.3% acetic acid	in 1% acetic acid	3
Shorr (1942) (Koss, 1979)			0.5 + 1.0 ml glacial acetic acid	0.5			0.25		Fast Green, 0.075	0.5	1
Papanicolaou (1942) (original) EA 36	0.23			0.17			0.5 + 0.015 PTA		Light Green, 0.23		2

with a modification of Masson's trichrome connective tissue stain. This was brought to Papanicolaou's attention by a colleague (Miss Fuller) and led to a number of experiments, including many combinations of acid dyes, some containing additional acids in the form of phosphotungstic acid (PTA) or phosphomolybdic acid (PMA) as in Masson's original stain (Papanicolaou, 1933, 1941). He also combined Eosin and Water Blue in one single solution to which he added PMA. To this mixture he added Orange G and Acid Fuchsin and thus achieved a variety of shades of pink, orange and red in the cytoplasm. Whether these different shades were caused by the differing amounts of the various dyes used, or by the varying amounts of acid added, was not explained.

The combinations of acid dyes which were originally recommended by Papanicolaou (1942) and further formulated in his later work (1954) are the Eosin–Light Green mixtures which are used at present and to which PTA is always added. Table 6.2 shows the many modifications of Papanicolaou's cytoplasm stain which can be found in literature. The list of authors is by no means complete and there may be many more cytologists who prefer their own composition of EA.

Essentially EA is a mixture of Eosin Y and Light Green to which Orange G, Bismarck Brown, phosphotungstic acid and sometimes lithium carbonate are added. The number of EA (e.g. 36, 50 or 65) is a code for the proportions in which the various components are present. Thus the differential staining of superficial and intermediate cells in which the colour of the cytoplasm ranges from red to blue or turquoise depends on the amount of dyes added and on the amount of acid.

6.3.2 Components of Papanicolaou's Stain and their Function

6.3.2.1 Phosphotungstic Acid and Phosphomolybdic Acid

In his paper 'A contribution to staining methods', Mallory (1900) introduced a differential method to dye connective tissue sections in the following way. First, he dyed the sections with Acid Fuchsin, then

Table 6.2   Some modifications of Papanicolaou's polychrome counterstain

	Connective tissue cytoplasm stains (g/100 ml)				
	Eosin Y	*PTA*	*Orange G*	*Fast Green or Light Green Y*	*Number of staining baths*
Rakoff (1960) (original)	0.17			Light Green, 4.17	1
Entschev (1963) (original)	0.3	0.6		Light Green, 0.5	1
Dunton (1972) (original)	0.19	1.7	0.08	Light Green, 0.37	1
Gill (1977)	0.4	0.2	0.4 +0.015 PTA	Light Green, 0.03	2
Wittekind and Hilgarth (1979); Wittekind (1980, pers. comm.)	0.04	0.03 + 1 drop glacial acetic acid + 1 drop lithium carbonate		Fast Green, 0.005	
Pamihall, pH 6.2 (Pawlick, 1977)	0.25	0.2 + 1 drop lithium carbonate	0.01	0.06, EA 36 0.04, EA 65 0.03	1

placed them in a solution of phosphomolybdic acid (PMA) after which he dyed them in a mixture of Aniline Blue and oxalic acid. He attributed the following functions to the various components: the oxalic acid was added to intensify the Aniline Blue, and the Orange G was supposed to 'limit' the blue to the connective tissue. According to Mallory the PMA acts in two different ways. First, it intensifies the Acid Fuchsin and fixes it to certain histological sites while at the same time removing it from the connective tissue. Secondly, it slows down the action of the aniline Blue and prevents it from gradually staining everything else in addition to the connective tissue. Mallory did not explain exactly how PMA acted but stated merely what it did. He vaguely indicated that some form of physical action was involved, such as difference in penetration of all components or exclusion from penetration.

This idea was further developed by von Möllendorff (1924) in his theories on the mechanism of staining (see section 3.7.3). According to these theories, PMA does not merely act as an acid like hydrochloric or acetic acid. von Möllendorff demonstrated the different effects by staining sections with Eosin Y and treating them afterwards with solutions of hydrochloric acid, acetic acid or PMA, respectively. The two former solutions caused discolouring of the whole section whereas treatment with PMA left the bright red Eosin dye in some parts of the sections and showed a clear differentiation between different tissue components. Thus in some elements of the connective tissue sections, PMA acted as an acid replacing the Eosin, and in others, in which the Eosin was still present, it did not. von Möllendorff's *in vitro* experiments using gelatin films showed that PMA penetrates much more slowly than hydrochloric and acetic acids. In tissues, this slower diffusion rate would account for the effect of differential staining: in tissue elements with a compact structure, Eosin can easily penetrate and PMA would need a longer time to replace it. It was shown by von Möllendorff that the differentiating effect of PMA decreased if it was allowed to act longer. When tested with other dyes, PMA did not act in the same way, so von Möllendorff suggested another mode of action, namely that of exchange diffusion where the dye which is already settled is exchanged for more slowly penetrating PMA. In this model PMA is thought to act like a competing dye.

This latter point was taken up by Baker (1970). He was able to locate the position of PMA in connective tissue by 'dyeing' with PMA and then exposing the slides to light. PMA is not a dye, but exposure to light oxidises it to a lighter blue molybdenum oxide which can easily be seen in collagen, less in muscle and least of all in red blood corpuscles. This order reflects a decrease of permeability for dyes from collagen to erythrocytes. A similar distribution of dyes was obtained with Methylene Blue, which is also a slowly diffusing dye. If sections are treated with PMA after staining with a fast penetrating dye like Eosin or Acid Fuchsin, the former competes with the latter. This is in accordance with Mallory's description of the action of PMA. The theories on the role of acid in polychrome cytoplasmic stains all relate to PMA. Nowadays PTA is mostly used and we must assume that PTA, which is very similar to PMA, acts in a similar way.

Lillie (1977) did not completely exclude the possibility of an explanation of the role of PMA and PTA on a physical basis, but was more inclined to look for a chemical explanation. According to him the high hydroxyl content of collagen could present an adequate number of binding sites for the polymeric PMA and PTA.

That difference in penetration plays some part is supported by the observation that prolonged staining with polychrome stain containing PMA or PTA changes the staining result.

6.3.2.2 The Dyes

The three dyes which are active in Papanicolaou's cytoplasm stain are all acid, namely Eosin Y, Light Green and Orange G. They are competing dyes, and each has a different peak pH at which it stains most intensely (figure 6.1a–c). Thus the colour of the cytoplasm, when more than one of these dyes is used (as is the case in the Papanicolaou method), is in principle a combination of these colours: red and orange–from Eosin Y, green from Light Green and yellow from Orange G. Depending on the pH of the staining-baths and on the concentration of the dyes, the colour may then range from yellow to red to blue to turquoise to green (see figures 6.2a and b).

Eosin Y is a purely acid dye, binding mainly to proteins. It can penetrate dense structures, and is metachromatic, as demonstrated by Marshall *et al.* (1981) in their specrophotometric analysis of Romanowsky stains: the monomer has a red colour and the dimer is orange–red. Used on its own, it stains the cytoplasm of superficial squamous cells red and erythrocytes orange–red. However, nuclei, especially with condensed chromatin, are also stained. Thus pyknotic nuclei can take up Eosin. this is evident from their purplish colour in the Papanicolaou staining method. Nucleoli can also stain red with Eosin (see chapter 7).

Light Green is an acid dye also containing basic groups, which means it is amphoteric. Used on its own, it stains all components of the cell green. It stains predominantly the cytoplasm.

Orange G is a small acid dye which can penetrate dense structures even better than Eosin. It stains dense structures yellow. It was originally added to acidify the solution and thus to intensify the effect of the acid dyes in the EA. The necessity for using Orange G at all in the Papanicolaou method is disputed (Drijver and Boon, 1983*b*). Orange G competes mainly with Eosin Y, staining the more dense structures. Thus highy keratinised cells and erythrocytes stain orange to yellow. Many commercial Orange G solutions have a pH at which Orange G does not stain at all. In that way, it acts only as an acidifier, changing the colour pattern of the EA.

Bismarck Brown is a basic dye which was originally added to Papanicolaou's stain. The necessity for its presence is doubtful (see section 6.1). It is no longer added to most commercial Papanicolaou polychrome stains.

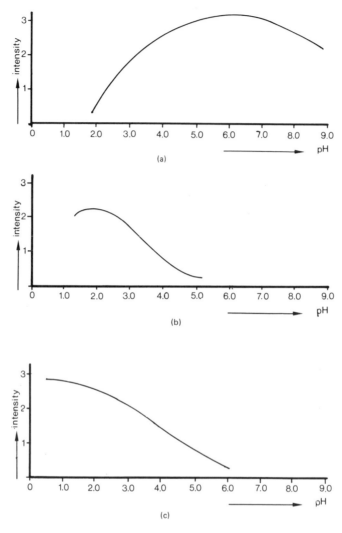

**Figure 6.1   Peak staining intensity of: (a) Eosin Y with HCl, (b) Light
Green Y with HCl, (c) Orange G with HCl.**

6.3.2.3 Effect of pH on the Staining Patterns of EA

The question arises as to what exactly determines
the differential staining and the colour of the
squamous cells when using the different EA solutions (Table 6.2). In section 6.3.2.1 it was made
clear that PTA not only acidifies but plays another
role as well.

The effects of pH changes of the EA and the role
of PTA were investigated by Drijver and Boon

(1983*b*). We stained a series of vaginal smears taken
on the same day of the menstrual cycle and all
containing intermediate and superficial cells with
EA solutions with different amounts of PTA so that
the pH ranged from 8.0 (when no acid was added)
to 1.0 (when 6% PTA was added). The cytoplasm
of intermediate cells was coloured differently according to the pH, ranging from red to blue to
turquoise and finally to green at the lowest pH (see
figure 6.3a). To check whether this was merely

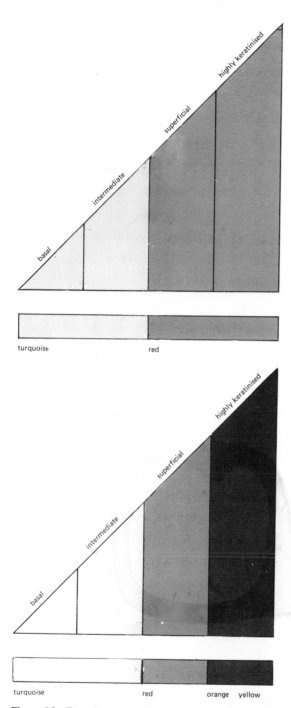

Figure 6.2 Top: the range of colours of the cytoplasm of different types of cells when stained with EA without Orange G (OG). Bottom: the same but now the cells are also stained with Orange G (pH 2.5).

caused by the pH, we made up EA solutions of similar pH but acidified with hydrochloric acid instead of PTA. The results can be seen in figure 6.3b). It is clear that the cytoplasm of intermediate cells stained blue (which is a mixture of Eosin and Light Green) at pH between 4 and 5 whereas EA acidified by PTA stained the cytoplasm of intermediate cells turquoise at those pH values.

These findings made further experiments necessary on the influence of acidity on Eosin and Light Green individually. The results are shown in figure 6.4. From the figure two things become clear. First, the intensity of both Eosin and Light Green is increased by the addition of more acid. Secondly, the intensifying effect of PTA on Eosin is less marked than the intensifying effect of PTA on Light Green. The intensity of Eosin is increased more by hydrochloric acid than by PTA. The result is that at lower pH Light Green gets a better chance to manifest itself, changing the colour pattern of the EA as shown in figure 6.3a.

Puchtler and Isler (1958), using histological sections, contend that PTA has a mordanting effect for Light Green. This would explain the marked effect of PTA on Light Green in our experiments. The acid groups of the PTA already form a bond with the basic groups of the proteins and to this is added the bonds in the basic dye group–PTA–basic protein group. If PTA is considered as a competing dye, it seems to compete more strongly with Eosin than with Light Green (see figure 6.4). In comparison with the acidifier hydrochloric acid, without any competing action with either Eosin or Light Green, PTA decreases the staining results of Eosin but not of Light Green. The overall picture is that of three competing 'dyes' (Eosin, Light Green and PTA), with the added mordanting effect of PTA on Light Green.

In histological sections the differentiating effect of staining with a combination of acid dyes is explained by their different rates of penetration into the tissue elements (Singer, 1952). This is caused by: (a) the difference in size of the dye molecules, and (b) the difference in density of the various tissue elements. In cytology also, the competition of Eosin and Light Green can be explained by the difference in size of the dye molecules, the Light Green being

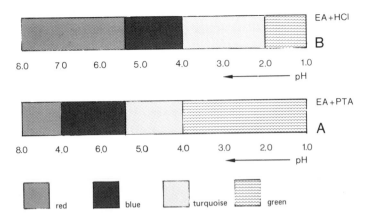

Figure 6.3 **The colour of the cytoplasm of intermediate cells in relation to pH of the EA solution. (A) EA acidified by phosphotungstic acid. (B) EA acidified by hydrochloric acid.**

the largest. The difference in density of the individual cells is the factor which determines the difference in penetration. Ascher *et al.* (1956) demonstrated more disulphide bonds (S − S) in the cytoplasm of superficial cells (with small nuclei) belonging to keratin precursors, and more sulphydryl groups (SH) in the proteins of the cytoplasm of intermediate cells (with larger nuclei, see figure 6.5). The disulphide bond gives the protein a tighter structure so that the superficial cell becomes less penetrable by a dye with a larger molecule, i.e. Light Green. The difference in rate of penetration between Light Green and Eosin is also visible in the inner cells of groupings: if the cell groupings are thick, the inner cells invariably stain red.

6.3.2.4 Effect of pH on the Staining Patterns of Orange G

Orange G is only effective around pH 2.5 (see figure 6.1c). It then stains erythrocytes and keratinised cells yellow. The more keratinised the cell, the more stain is found in the cytoplasm. Anucleated squames and the cytoplasm of well differentiated

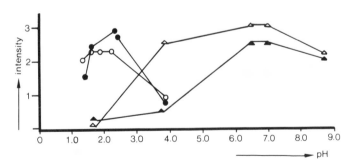

Figure 6.4 **The intensity of staining with Light Green Y (Series III & IV) and Eosin Y (Series V & VI) in relation to pH of the staining solution. ● = Series III (Light Green with PTA); ○ = Series IV (Light Green with HCl); ▲ = Series V (Eosin with PTA); △ = Series VI (Eosin with HCl).**

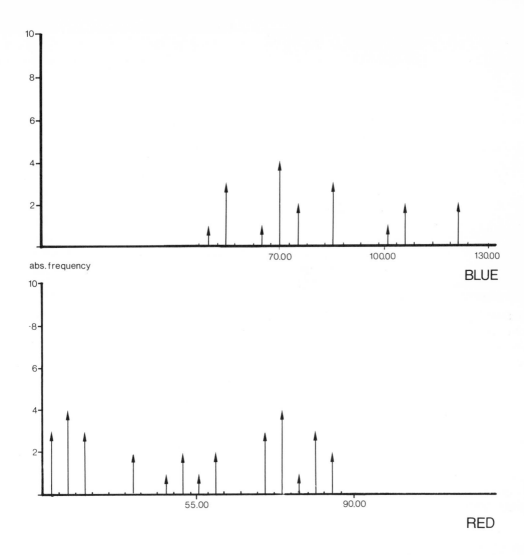

Figure 6.5 **Absolute frequency of nuclei of squamous cells with blue and red cytoplasm, of smears stained with Leiden–Papanicolaou method. Note that cells with small nuclei are predominantly red (superficial cells) and cells with larger nuclei predominantly blue (intermediate cells).**

squamous carcinoma cells are stained bright yellow. When Orange G is used on its own, the intensity of the colour indicates the degree of keratinisation of the cell.

When used in the Papanicolaou method prior to staining in EA, the staining result will depend on the pH of the Orange G and that of the EA. If the pH of the Orange G is over 3.5 and lower than 6.5, it will not have any colouring action on its own but will intensify the staining results of the EA. If the pH of the Orange G solution is 2.5, it will compete with the Eosin in the EA (see figure 6.2). When the pH of EA is most favourable for Eosin (that is, around 6.5), the action of Eosin is visible in the red

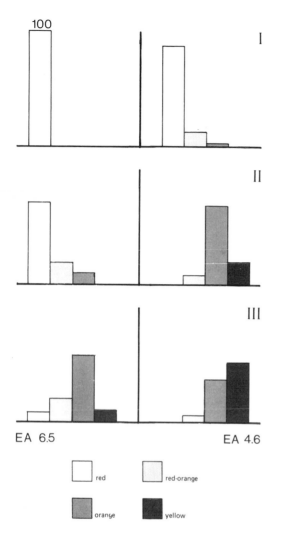

Figure 6.6 **Histograms of squamous cells stained with Orange G, pH 2.5, and EA pH 6.5 and EA pH 4.6. I, buccal smear; II, smear from skin; III, skin from highly keratinised sole of foot.**

The orange in Papanicolaou stained cells can be due either to the dimer of Eosin Y or to a combination of the monomer of Eosin Y (red) and of Orange G (yellow).

6.3.2.5 Effect of Fixation and Staining Times

The staining results of the polychrome Papanicolaou methods are dependent not only on the pH of the stains but also on the fixation of the cells. The cells should be wet fixed (see chapter 4). If this wet fixing is performed too slowly, the cells become air dried. The staining pattern of air-dried squamous cells differs significantly from that of wet-fixed cells (see section 4.4.1). Nucleoli in air-dried cells do not take up Eosin, so these will *not* stain red.

The staining time also plays a part in the results. For wet-fixed squamous cells in cervical smears, the results for EA staining times between 2 and 4 min (normal) do not differ too much (see figure 4.3). From the above it is clear that, when normal staining times are employed, an air-dried smear contains a much higher proportion of red staining cells in comparison with the parallel wet-fixed smear. For very refined hormonal studies, one should really stain wet-fixed smears for *60 min* in Leiden EA: the results are then highly reliable, and the staining patterns pleasing to the eye. When longer staining times are used, the stain in the blue–green cells is rather pale. The Shorr staining technique is less sensitive to variations in staining times, and is therefore easier to apply in hormonal studies (Pundel, 1950).

6.3.2.6 Effect of Specimen-dependent Factors on the Staining Patterns

As has become clear, the thickness of cell groupings is an additional important factor in the staining results. The smaller dye molecules penetrate best, thus thick cell groupings will stain orange or red depending on the method used. Furthermore, the 'background' of the smear is important. Cells close to blood and fibrin have a tendency to take up the smaller dye molecules, and near mucus the larger ones.

staining of the superficial squamous cells. In that case, the erythrocytes stain orange, and highly keratinised cells (e.g. from squamous cell carcinoma of condyloma acuminata) stain orange to yellow. When the pH of the EA is slightly unfavourable for Eosin (that is, around 4.6), the superficial squamous cells also stain orange (see Atlas, plate 4.1 and figure 6.6). The highly keratinised cells are then yellow (figure 6.6).

6.3.2.7 Manipulating the Papanicolaou Staining Methods

In our opinion it is not correct to speak of 'the' Papanicolaou stain. It would be more appropriate to speak of 'a' Papanicolaou stain and always to mention the pH of the EA mixture and the length of time of staining, and the use of Orange G and its pH. These factors are of great importance, as can be seen in our photographs. Depending on the material, and to suit the taste of the individual cytologist, the Papanicolaou stain can be manipulated by monitoring the pH of the EA and its dye concentrations, and of the Orange G, if used. In this context it is important to note that if the pH of a commercial EA is altered, the results might differ somewhat from the results of a change in pH of laboratory-produced EA, as described in appendix 2 due to differences in dye concentrations.

If the cytologist likes the cytoplasm to be turquoise, the pH of the EA should be around 4.5. If a blue cytoplasm is desired, the pH should be raised to around 6.5.

Orange G should only be used at pH 2.5. The EA should be raised to 6.5 if Orange G is included: it then colours highly keratinised cells orange–yellow. In thick smears and in smears containing much blood, it is an unpleasant staining method because thick cell groupings and erythrocytes show a glaring orange colour.

By changing the pH of the EA, the method can, using Orange G at pH 2.5, visualise the different stages of keratinisation of squamous cells. This is illustrated in figure 6.6. At pH 6.5 of the EA, the buccal smear does not contain any orange cells, and 92% of the anucleated squames of the maximally keratinised epithelium of the foot stain orange–yellow. At pH 4.6 of the EA, 16% of the squamous cells stain orange, and all cells from the callus of the foot. Other, less keratinised, skin (thumb) takes an in-between position.

If one wants to see nucleoli red in blue nuclei, the EA should be around 4.5 (see chapter 7). At higher pHs, only large nucleoli in malignant cells stain red (see Atlas, plate 5).

An individual smear can be destained and res-

tained at other pH values in order, for instance, to visualise abnormal keratinisation. It is clear that manipulating the colour patterns of the Papanicolaou stain will enhance cytodiagnosis.

6.3.3 Shorr's Polychrome Stain for Hormonal Cytology

For hormonal studies Shorr developed a staining technique which was more rapid than Papanicolaou's. He published the changes that took place in response to ovarian hormones in 1942. The principle of Shorr's staining method is the same in Papanicolaou's: the cytoplasmic stain contains competing dyes with different rates of penetration. However, Shorr's method is less sensitive for variations in pH and staining times.

In his method, the counterstains Biebrich Scarlet, Orange G and Fast Green are put in one solution. The change from Eosin to Biebrich Scarlet and from Light Green to Fast Green was made for purely practical reasons since at that time Eosin and Light Green were difficult to get hold of in the United States as they had to be imported from Europe. As acidifiers, phosphotungstic acid and phosphomolybdic acid are added. The superficial cells in cervical smears stain a brilliant orange–red. The intermediate cells stain green. Nucleoli are burgundy red. As in Papanicolaou's method, the cells should be wet fixed.

6.3.4 Rakoff's Polychrome Stain for Hormonal Cytology

Rakoff developed a staining method that can be applied on unfixed cells, and can be used as a first method for hormonal cytology by the gynaecologist. It is a mixture of the two cytoplasmic dyes Eosin Y and Light Green Y in 96% alcohol, and provides the possibility of a quick first diagnosis. Lugol's solution (appendix 3) may also be added for demonstration of glycogen.

6.4 Histochemical Staining Methods

The role of histochemical staining in the cytology laboratory can be of the same importance as in the histology laboratory (Sachdeva and Kline, 1981). In general, the staining times are shorter. Certain histochemical staining methods can be used in differential diagnoses, and some in the study of circadian or monthly rhythms (Rietveld and Boon, 1981).

The most frequently used histochemical staining method in our laboratory is the PAS reaction. Polysaccharides can be visualised with it, as is discussed in chapter 3. In this method, the '1,2-glycol' group of polysaccharides is oxidised with periodic acid. The aldehyde groups thus liberated give a positive Schiff reaction.

PAS positive material is stained red. The PAS reaction is positive for mucin, mucoproteins, hyaluronic acid, colloids and glycogen. It can be performed on air-dried or wet-fixed slides. Ethanol does not fix carbohydrates but precipitates them unchanged. Air-dried smears should be post-fixed for 15 min in 90% ethyl alcohol. The PAS reaction can also be performed on destained slides, but since the red colour differs from the colours in the Romanowsky–Giemsa and the Papanicolaou methods, the PAS reaction can also be performed on cells which are already stained in that way. In contrast to what occurs in formalin-fixed embedded cells, PAS-positive material in cytological preparations is not dislocated in the cells.

When the cells are treated with amylase prior to staining, all polysaccharides and substances containing polysaccharides stain negative with the PAS reaction. This is due to the fact that α-amylase releases the α-1,4-glycoside bonds, leaving only disaccharides and trisaccharides intact, and β-amylase splits glycogen (Lehninger, 1975). Those substances that are negative with the PAS reaction after amylase treatment are called diastase resistant. Table 6.3 shows that it is not only glycogen that disappears after amylase treatment. Bulmer (1959) introduced the use of the aldehyde blocking dimedone (5,5-dimethylcyclohexane-1,3-dione). After oxidation with periodic acid, in which aldehydes are formed, the material was incubated for a short period of time with dimedone. After this treatment, polysaccharides, mucoproteins and mucin are still PAS-positive. Bulmer concluded that the best way to perform this method is to incubate in 5% dimedone alcoholic solution at 60°C. By extracting the values of staining without pre-treatment with amylase, with pre-treatment with amylase, and with dimedone incubation, the percentages of cells with respectively glycogen and mucin (and mucoproteins) in a smear can be calculated (Veeken, 1983).

The best method for preserving glycogen in air-dried cells is to post-fix the smears in 80% methanol according to the method of Gabe (1962).

Table 6.3   PAS-positive reaction visualising various components

Without pre-treatment	Pre-treatment with amylase	Dimedone treatment
Polysaccharides	—	Polysaccharides
Mucoproteins	Mucoproteins	Mucoproteins
Glycoproteins	Glycoproteins	—
Glycolipids	—	—
Phospholipids	—	—
Mucin	Mucin	Mucin

Table 6.4   PAS—reaction, staining patterns

Cells	Staining pattern
Mesonephroma	Diffuse
Mesothelioma	Predominantly outer zone of cytoplasm
Signet cell carcinoma of the stomach	One big vacuole
Dysgerminoma	Diffusely granular
Mucus-producing adenocarcinomas	Multiple vacuoles of varying sizes
Cryptococcus neoformans	Capsule of organisms stains diffusely
Cells from breast carcinomas	PAS-positive dot in intra-cytoplasmic inclusions
Benign squamous cells in vaginal smears	Diffusely intense in patients with good response to treatment of cervix carcinoma
Benign squamous cells of buccal mucosa	Diffuse; intensity related to circadian rhythm
Benign squamous cells in vaginal smears	Diffuse, dependent on hormonal conditions

In diagnostic cytology it is important not only to ascertain whether a cell stains positive with the PAS reaction, but also to evaluate the staining *pattern*. As is clear from table 6.4, different cell types have different staining patterns. For instance, in malignant mesothelioma, the PAS-positive material is mainly situated in the 'blebs' and in adenocarcinoma cells in large vacuoles (see Atlas, plate 21.1) (Boon *et al.*, 1982a, 1984b).

The characteristic distribution of PAS-positive material in the cytoplasm can aid the cytologist in identifying the mesothelial origin of the cells. In our series (Boon *et al.*, 1984b) there was one case of metastatic carcinoma in which the malignant cells also displayed blebs: in this case the PAS stain was of additional value as the blebs were PAS-negative. The PAS reaction is also very valuable in the detection of *Cryptococcus neoformans*.

The PAS method can also be used in the distinction of breast carcinomas: Johansen and Thuen (1981) reported that 62% of the carcinomas and only 3% of the benign lesions had PAS-positive material in the cytoplasm of cells of imprint smears. This PAS-positive material is seen as a dot in an intracytoplasmic inclusion. The PAS method can also provide important information in the study of cervical carcinoma (Das and Chowdhury, 1981), and in hormonal studies (Chowdhury and Chowdhury, 1981).

The amylase treatment prior to the PAS reaction can be used to identify mucin (D-PAS). In two of our cases, using this method the malignant carcinoma cells could only be identified in pleural fluid, because the nucleus was flattened by the large D-PAS-positive vacuole and thus its malignant nature was not apparent (see figure 6.7).

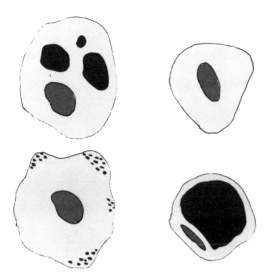

Figure 6.7   PAS staining patterns of malignant mesothelial cell and adenocarcinoma cell (signet-ring type). PAS-positive areas black. Upper left: adenocarcinoma cell; upper right: no PAS-positive material; lower left; mesothelial cell with PAS-positive blebs: lower right: adenocarcinoma cell, signet-ring type.

The Alcian Blue stain can also be performed on air-dried or wet-fixed slides. One-fifth of our metastatic carcinomas stained positive (table 6.5), and the

positive staining was always in distinct vacuoles. In contrast, the positive staining of the malignant mesothelioma cells was invariably located in the area of the villi (see Atlas, plate 19).

For fat staining it is important to evaluate the distribution of the positive material. Different cell types have different patterns (see table 6.6). Mesothelial cells have a characteristic distribution of lipid vacuoles, particularly in cells maturing while floating in the coelomic fluid (see chapter 2). Sex hormones influence the size and distribution of lipid granules in vaginal cells (Lahiri and Chowdhury, 1981), and fluctuate during the menstrual cycle (Bibbo *et al.*, 1969). The fat staining must be performed on air-dried slides within one week of sampling. Even when these precautions are taken, some lipid vacuoles will have lost their contents (see Atlas, plate 20.4).

We prefer the ORO stain for visualising fat.

6.5 Staining Pigments

Clinically it can be very important to identify haemosiderin pigment in macrophages in sputum, or melanin in malignant cells. Light-microscopical-

Table 6.5   Staining patterns in cells from malignant mesothelioma and metastatic carcinoma exfoliated in pleural fluid

	Malignant mesothelioma		Metastatic carcinoma	
	Vacuoles	*Cytoplasm outside vacuoles*	*Vacuoles*	*Cytoplasm outside vacuoles*
ORO (Oil Red O)	17	0	2	0
PAS	0	17	5	4
D-PAS	0	0	5	0
Alcian Blue	0	8	5	5
N	17	17	25	25

ly, haemosiderin has a refractive brown–orange appearance in the Papanicolaou method. The smear can be destained and restained with one of the iron stains described in appendix 2. The Schmorl stain, although not specific for melanin, can be used if clinically a malignant melanoma is suspected or when brownish particles are seen in the cytoplasm. This method can also be performed on destained slides (see Atlas, plate 1.)

6.6 Staining Fibrous Components

Clinically it can be of importance to show that tumour cells produce fibres. For this reason a silver impregnation stain like Gomorri's can be used on cytological material. Lopez Cardozo (1977) places great importance on the aspect of the fibres in the distinction between the different entities. (See Atlas, plates 22 and 23.)

Staining of the Nucleolus

7.1 Introduction

In section 1.1.4 we saw that the nucleolus comprises a fibrillar and a granular zone, each containing RNA and ribonucleoproteins. There is some DNA present in the fibrillar (inner) zone. The ribonucleoproteins are non-histones, which means they are acid. Both Papanicolaou's and Romanowsky–Giemsa's staining methods show the presence, form and number of nucleoli quite clearly. In Papanicolaou's method they are either red or blue (depending on the pH of the fixative and of the staining baths). In Romanowsky–Giemsa's they are all shades of blue.

It may be desirable to stain the nucleolus only or even just a part of it. First, we will look at a number of specific nucleolar staining methods. We will not mention all of the methods that have been described in recent years. We have only handpicked those that represent a particular feature of the nucleolus. A detailed description of nucleolar staining methods is provided by Busch and Smetana (1970). We shall be looking at the following staining possibilities:

(1) a basic dye together with a metal salt to show single-stranded RNA;
(2) a basic fluorescent dye at acid pH to show nucleolar RNA with a DNA counterstain;
(3) a basic dye after DNA digestion and followed by molybdate treatment to show the fibrillar centre of nucleoli (nucleolini);
(4) an acid dye after mordanting with phosphomolybdic acid to show proteins in the interphase nucleoli;

(5) an acidified acid dye which is easy to apply on smears;
(6) staining with silver for research purposes.

Secondly, we will discuss the staining of nucleoli in the Papanicolaou and Romanowsky–Giemsa methods.

7.2 Some Nucleolar Staining Methods

7.2.1 Cuprolinic Blue for the Selective Staining of Single-stranded RNA

Cuprolinic Blue is related to Alcian Blue which is known as a typical glycoprotein dye. CB can also bond with negatively charged polyanions. It stains both DNA and RNA but with the difference that the bond between CB and DNA can easily be broken down by addition of cations in the form of metals. If metal ions are added, the salt complexes of CB and DNA are split up and those between CB and RNA remain. With this staining method (Mendelson *et al.*, 1983) it is possible to make relative RNA measurements by determining the absorbance at 635 nm.

7.2.2 Fluorescent Staining of RNA with Acridine Orange

This is another way of staining nucleolar RNA selectively. Acridine Orange is a basic dye which does not stain DNA or the ribonucleoproteins, but stains RNA selectively orange to red. For contrast DNA can be stained by Methyl Green. The staining solution should have a pH of 4.9.

7.2.3 Toluidine Blue for the Selective Staining of Nucleolini

The azure dyes can be used as selective nucleolar stains under certain conditions. By using a special technique, Love *et al.* (1973) were even able to stain the nucleolini differentially. The method involves blocking of the amino groups of the nucleoproteins by trichloracetic acid, DNAase digestion, staining with Toluidine Blue and after-treatment with ammonium molybdate. The latter is a delicate process. It should not last longer than 7 min and the solvent should be double glass- or quartz-distilled water. The molybdate is supposed to induce polymerisation of the dye (see section 3.9.3) so that the nucleolini become stained metachromatically. It is suggested that it is the RNA in the nucleolini which is stained in this method (Pearse, 1980). Others use Toluidine Blue as a straightforward nucleolar dye without DNAase digestion or molybdate after-treatment. All methods have in common that the solvent should have a low pH of between 3 and 4.

7.2.4 Staining with the Acid Dye Pontacyl Dark Green B

Bedrick (1970) tested a number of acid dyes and found acidified Pontacyl Dark Green B very acceptable as a nucleolar dye. Unfortunately, no pH is mentioned. The method involves no further pre- or post-treatment and seems quite straightforward. It has been tested on tissue sections but would probably serve for cell samples as well.

7.2.5 Staining with the Acid Dye Amido Black 10B

In this method cells are first mordanted with phosphomolybdic acid and then dyed. Again no mention is made of the pH of the dye solution used. The method gives bright blue nucleoli also in cell smears and is easy to apply (Wood and Green, 1958; Mundkur and Greenwood, 1968).

7.2.6 Staining with Silver

Busch *et al.* (1982) have done some very interesting work on ribonucleolar proteins. They were able to separate and identify a number of these in the nucleolus. The fibrillar centre of the nucleolus is Ag-positive. With this method it is possible to trace ribonucleolar proteins. The amount of these is greater in nucleoli of tumour cells than in nucleoli of normal cells. Obviously this is not a suitable method for routine techniques but it is included here since it gives a very valuable insight into the fact that nucleolar proteins of malignant cells may differ from those of benign cells.

7.3 Staining of Nucleoli in the Papanicolaou Methods

As is discussed in chapter 6, all three cytoplasmic dyes used in the Papanicolaou method, that is, Eosin Y, Light Green Y and Orange G, can stain the nucleolus. So what colour will it have in the Papanicolaou method? Evidently, the colour range is from green to blue to red to orange, and will depend on which one of the competing dyes will have the best chance of taking effect. The chances of the cytoplasmic dyes can be manipulated, as is shown in chapter 6, by changing the pH of the colour baths. Thus, in a given cell population, the colour can be orange when the pH of the Orange G is 2.5, and the pH of the EA is 'unfavourable' for Eosin at around 3.0. The colour can be green if no Orange G is used and the EA has a very low pH, and red when the pH of the EA is around 4.7 or the Eosin concentration is still high enough. (See Atlas, plate 5.)

In addition, the pH of the fixative plays an important role in the staining of the nucleolus. If the fixative is too basic, it is difficult to stain the nucleoli differentially from the chromatin.

Last but not least, Haematoxylin stains the nucleolus blue. If the acid Orange G bath follows the Haematoxylin bath, the additional staining of the nucleolus by Haematoxylin is reduced. For some reason, nucleoli in air-dried cells that are not treated with prolonged incubation with saline formalin invariably stain blue. The resulting colour of the nucleolus is always a combination of staining with Haematoxylin (blue) and the three cytoplasmic stains. If one wants to stain the nucleus differentially from the chromatin, one should try to reduce the Haematoxylin staining of it as much as possible by optimal differentiation and thorough rinsing.

Mouriquand *et al.* (1981) observed that the colour of nucleoli in a given Papanicolaou staining method can be useful in distinguishing benign from malignant cells. In their method, the nucleoli of benign cells were blue and those of malignant cells red. These authors, however, did not publish the pH values of their fixative and staining baths. We have confirmed their observation that in some Papanicolaou methods there is a difference in colour between the nucleoli of benign and malignant cells. Further investigation is needed, using pH-controlled fixation and staining, to study whether these phenomena can be applied in routine diagnostic cytology.

The visibility of the nucleolus in the Papanicolaou method is dependent not only on its colour, but in addition on the chromatin pattern. This is summarised in table 7.1. Air-dried nucleoli do not stain with Eosin, and thus are not red.

Table 7.1   Visibility of nucleolus in the Papanicolaou method

Clearly visible	*Not clearly visible*
Nucleolus red	Nucleolus blue or green
Chromatin open	Chromatin dense
Chromatin fine	Chromatin clumped
Perinucleolar halo	No perinucleolar halo

7.4 Staining of Nucleoli in the Romanowsky – Giemsa Methods

In the Romanowsky–Giemsa (R–G) methods, nucleoli of well stained cells in a monolayer are differentially blue in purple chromatin (see table 7.2). However, when the Romanowsky effect is suboptimal or absent, the chromatin is also blue (see chapter 9). It is then difficult or impossible to see the blue nucleolus, especially when the nuclear staining is intense. Longer staining times are then needed to produce the purple staining of chromatin.

Table 7.2   Visibility of nucleoli in the Romanowsky – Giemsa method

Clearly visible	*Not clearly visible*
Nucleolus light blue	Nucleolus dark blue
Chromatin open	Chromatin dense
Chromatin fine	Chromatin in clumps
Cells in monolayer	Cells in cluster

In the presence of formalin (vapour) the nucleoli of fresh cells are changed such that they will also tend to have a blue nuclear stain (see section 9.3.5). In these, very *short* staining times are needed to visualise the dark blue staining nucleoli in faintly blue staining nuclei.

The kind of fixation has its influence on the colour of the nucleolus after R–G staining. This can be seen in figure 4.4.

Vital Staining Methods and Quick Stains

8.1 Introduction

In certain situations the cytologist wants to have a first impression of the cell sample taken, or a preliminary diagnosis is desired by the patient and/ or clinician. The first is important in aspiration cytology, to check whether the lesion or organ is aspirated and whether the sample contains diagnostic cells. A preliminary diagnosis of an aspirate is necessary in order to plan further investigation or an operation. In the latter circumstance, it takes the place, or is applied in conjunction with, frozensection histology (Pickren and Burke, 1963). In both instances, it is important that the cell sample be ready for microscopy soon after it is taken. Thus short staining and, if fixing is needed, short fixing times are extremely important. Since a refined diagnosis is not needed here, a compromise in visual quality, in order to meet the desired time limits, is acceptable.

These fast methods can be applied on live cells. There is no differential staining, and, in most

methods, no fixation, thus the slides cannot be kept. In addition, some fast methods have been developed for air-dried and wet-fixed slides. In these, some differentiation is achieved, although this is sub-optimal; these slides can be kept.

In addition, many clinicians use phase contrast microscopy, which does not need any staining and will not be discussed in this book.

8.2 Vital Staining Techniques

Ehrlich introduced Methylene Blue in 1877 as a vital dye for nerve axons and applied it in his later work on cytological material (see Ehrlich, 1956). Around the turn of the century most vital staining was performed to study the chemical composition of the heterogeneity of cell inclusions. These staining methods can also be applied in cytodiagnostics. Baker wrote a very interesting chapter on cytological vital staining techniques in his book *Cytological Technique* (Baker, 1950).

8.2.1 Mechanisms

Staining of vital cells can be achieved (table 8.1) by active uptake of coloured material (not necessarily a dye), or because cells passively allow the dyes to infiltrate (see next section). In diagnostic cytology, only the latter mechanism plays an important role in vital staining. The vital stains used in diagnostic work are as follows:

Table 8.1 Mechanisms of vital staining

Mechanism	Results
Cells actively take up coloured particles	Coloured ingested particles visible in cytoplasm
Cells allow dyes to infiltrate	Colouring of cellular constituents

Azure A and B
Methylene Blue
Janus Green
Bismarck Brown
Brilliant Cresyl Blue
Toluidine Blue

Vital stains can be orthochromatic, but many have metachromatic properties (the thiazines and Janus Green). An object that is a chromotrope in the fixed cell is frequently a chromotrope in the living cell as well.

8.2.2 Staining Patterns

Of the living cells, certain vacuoles, lipid globules and mitochondria are coloured first and then the nucleolus. Slowly the nucleus and then the cytoplasm become diffusely coloured. A chromatin pattern, as in fixed cells, is not visible. The fact that chromatin itself is not vitally stained can be attributed to the fact that in the living, unfixed cell, the DNA is combined with protein and therefore its phosphoric groups cannot react with cationic dyes (see section 1.1).

In general one should not expect that the morphological image of a fixed stained cell is the same as that of a vitally stained (unfixed) cell. In the latter not only is the chromatin pattern different, but in addition all kinds of vacuoles and inclusions, unseen in the fixed cell, are visualised. It is only the nucleolar/nuclear ratios and the nuclear/cytoplasmic ratios that are not fundamentally different.

When a living cell is left in the stain for a period longer than 1 h, its appearance begins to change. Therefore light-microscopical examination should be performed within the hour. In general, dyes kill cells. Vital stains should be used at low concentrations (0.01%) to postpone cell death.

The vital dyes have different preferential staining patterns. Brilliant Crystal Blue, for instance, is a good stain for the nucleolus and Janus Green B for the mitochondria.

8.2.3 Qualities of Vital Stains

8.2.3.1 Penetration

Overton (1890), studying the penetration of dyes in living cells, came to the conclusion that cationic dyes penetrate better than most anionic ones. He contended that this was due to the fact that they are soluble in phospholipids and therefore can pass the cell membrane.

Ionised Methylene Blue hardly penetrates through intact cell membranes. The dye can be reduced to its leucobase and it is possible that this occurs in the vicinity of the living cells (Baker, 1963). If this is true, the colourless leucobase passes through the cell membrane, and in the cell it is oxidised to the coloured chromophore. Penetration is pH dependent: acidity makes the cell membrane more permeable to the leucobase. At very low pH (<5), new cations of the leucobase are produced, which cannot easily penetrate.

8.2.3.2 Differential Staining

All vital dyes stain the cytoplasm and nucleus, and therefore these are coloured more strongly than the surrounding stain solution. The staining is achieved by flocculation of the dye *in* the cell, but in addition dyeing in the strict sense occurs (Baker, 1963, and section 3.9.3).

Except for metachromasia, in most vital staining methods there is no differential staining of cytoplasm and nucleus. With Methylene Blue, both are stained blue, but in most cells the nuclei stain darker than the cytoplasm and thus their sizes can be analysed by light microscopy.

8.2.4 Vital Staining of Cell Preparations

Vital staining can be performed on all kinds of cell preparation; aspirates, brushes, smears, sediments or touch preparations from surgical biopsies. The colouring should be done as quickly as possible after sampling.

Saline solutions of vital dyes cannot be kept for a long time, because they all flocculate easily. It is best to keep a stock solution of 0.5–1% in distilled water and to dilute this with the physiological saline solution first before use. It is also possible to dissolve the dye directly in the tissue fluid of, for instance, the aspirate, or of the concentrated sediment. Then the dye is dissolved in 95% ethyl alcohol and the solution is spread over the glass slide. The sediment or aspirate is brought on to the slide after the ethyl alcohol is evaporated. Then the dye can dissolve in the fluid and colour the living cells.

Light microscopical study should take place within 1 h (see section 8.2.2).

8.3 Quick Staining Techniques

Quick staining techniques are employed to evaluate the sample when fine needle aspiration is performed. The adequacy of the sample can be assessed before releasing the patient. If the quick method is applied, the accuracy of fine needle aspiration cytology can be improved from 86 to 97% (Pak *et al.*, 1981). In addition it is used in the operating room (Mavec, 1967; Godwin, 1976). In all quick staining methods, the fixing times are shortened. The fine chromatin details are therefore not visualised, but cytoplasmic, nuclear and nucleolar size can be assessed.

The quick staining methods can be divided into those for the Romanowsky–Giemsa method, for the Papanicolaou method and for the Haematoxylin–Eosin method. Examples of each are given in appendix 2.

For the Romanowsky–Giemsa method commercial staining sets are also available. Using these, air-dried smears can be stained within 15 s. Some of these staining solutions contain other ingredients than are found in the original Giemsa method, such as Fast Green in 'Dif-quick' (Merz and Dade, Switzerland).

In general, the dye concentrations in the quick staining methods are higher than those normally used. Another method to reduce fixing and staining times of slides is to use solutions at elevated temperatures (around 35°C), because chemical reactions require roughly half the time when the temperature is 10°C higher.

Neutral Stains: the Romanowsky–Giemsa Methods

9.1 Introduction

In previous chapters we have dealt with 'acid' and 'basic' dyes (see chapters 5 and 6). When these acid dyes are combined with other acid dyes, there is no chemical interaction. The same is true for a combination of basic dyes. However, a different situation arises when an acid dye and a basic dye are put in the same solution. In that case the coloured anion and the coloured cation may combine, forming a coloured salt. Such a solution is described as neutral. The salt as such can react with certain cell components. In addition, the acid component and the basic component of the neutral stain can react with other cell elements. Thus all three, the salt, the cation and the anion, can colour different cell components. Because of this threefold action, a neutral stain can produce a very useful differentiation of cell components in one staining bath.

A combination of the basic thiazine dyes and the acid dye Eosin Y is widely used in routine cytology. Many names are attached to these types of stain: for historical reasons we will use the term Romanowsky–Giemsa staining methods.

9.2 A Short History of the Development of the Romanowsky–Giemsa Stains

The development of the Romanowsky–Giemsa stains is closely linked to that of other neutral stains in histology and cytology (see table 9.1). Ranvier's *Carmine Picrique* can be considered as one of the first 'neutral' stains. In it a saturated aqueous solution of picric acid is combined with a saturated aqueous solution of ammonium carminate. Carminic acid is an amphoteric dye that will act as a basic dye in acid solution (see section 3.12). The combination of picric acid and ammonium carminate results in the formation of Carmine Picrate. Whether Ranvier's stain was truly a neutral stain as defined in the previous section has never been confirmed, but the basic idea was there. The further development of neutral stains was mainly focused on finding a satisfactory stain for blood smears. For that purpose, Ehrlich made a combination of Methyl Green (a basic dye), Acid Fuchsin and Orange G (both acid dyes). This combination became known as Ehrlich's triacid dye. He succeeded in staining blood smears differentially: nuclei of leucocytes stained green and erythrocytes stained orange. The granules of polymorph leucocytes took up both dye cation and anion and were therefore called neutrophil leucocytes. Chemically speaking, the name triacid stain is incorrect, but it became known as such. It is still used as a tissue stain rather than as a stain for blood smears for which it was originally developed.

Table 9.1 shows how other investigators have worked along the same lines in the development of neutral stains in histology. However, the need for a satisfying blood stain remained, with the particular aim in mind of finding a stain that visualised malaria parasites in blood smears. In 1891, two Russian scientists, Romanowsky and Malachowski, made a combination of the basic thiazine dye

Table 9.1 History of blood stains. (Unless otherwise mentioned, these data are taken from *Conn's Biological Stains* (Lillie, 1977).)

Author and relevant literature	Dyes	Method of preparation	Use	Results
1875 Ranvier (Carmine Piqrique (Baker, 1963, p. 262)	Sat. aq. sol. picric acid; sat. aq. sol. ammonium carminate	Mix solutions, evaporate precipitate in distilled water	General histology	—
1879 Ehrlich (Lillie, 1977)	4.5% Methyl Green, sat. (20%); Acid Fuchsin; sat. Orange G	Add Meth. Blue solution to Acid Fuchsin solution; add excess Acid Fuchsin till precipitate is dissolved	Blood films	Erythrocytes red, eosinophil granules red, nuclei deep blue, neutrophil granules dark violet
1882 Ehrlich's triacid (Lillie, 1977, p. 492)	Orange G; Acid Fuchsin; Methyl Green	—	Heat-fixed blood films	—
1885 Bernthsen (Baker, 1963, p. 265; Lillie, 1977, p. 494)	Methylene Violet; Methylene Azure, first mention	Polychroming of Methylene Blue with AgO_2	—	—
1886 Babes (Lillie, 1977, p. 492)	Same as Ehrlich	—	Tissues	—
1888 Heidenhain (Lillie, 1977, p. 492)	Same as Ehrlich	—	Intestinal mucosa	—
1888 Chenzinsky (Lillie, 1977)	Sat. Methylene Blue (4.5%) 25 ml (= excess); water, 25 ml; 0.5% Eosin Y in 60% alc., 50 ml	—	Blood films	Erythrocytes pink, plasmodia blue, leucocyte nuclei blue
1890 Plehn (Lillie, 1978)	Sat. aq. Methylene Blue (4.5%), 60 ml; 0.5% Eosin Y, 20 ml; distilled water, 40 ml; 20% KOH, 12 drops	Add KOH before using	Blood films	Erythrocytes pink, eosinophil granules pink, leucocyte nuclei deep blue, plasmodia blue, neutrophil granules not mentioned

Table 9.1 — *continued*

Author and relevant literature	Dyes	Method of preparation	Use	Results
1891 Malachowski (Lillie, 1978)	Eosin Y, Borax Methylene Blue	Same as Plehn	Malaria parasites in blood films	—
1891 Romanowsky (Baker, 1963, pp. 264, 266; Lillie, 1978)	Basically Chenzinsky but without alkali; sat. aq. Meth. Blue 2 parts, 1% aq. Eosin Y 5 parts	Mixture left to ripen for 2 months; *no alkali;* mix before use	Malaria parasites	Nuclei red–purple, parasite chromatin red–purple, cytoplasm Prussian blue
1891 Unna (Baker, 1963, p. 267)	Methylene Blue, potassium carbonate (quantities unknown)	Polychroming of Methylene Blue with potassium carbonate	—	—
1898 Nocht (Lillie, 1977, 1979, p. 494)	Unna's polychrome Methylene Blue, time of polychroming longer with less alkali: neutralised Unna 1 ml; sat. Methylene Blue till violet turns to pure deep blue, 1% aq. Eosin Y 0.2 ml	Dilute with water; mix immediately before use.	Malaria parasites in blood films	Unstable
1899 Jenner (Baker, 1963; Lillie, 1977, p. 494)	Sol. Methylene Blue (un-polychromed); Eosin Y	Precipitate collected and re-dissolved in methanol	—	No nuclear stain, erythrocytes bright red, eosinophil granules deep red, nuclei of leucocytes blue, basiphil cytoplasm of lymphocytes blue, basiphil cytoplasm of malarial parasites blue
1901 Reuter (in Germany) (Baker, 1963, p. 267)	Nocht's combination of Eosin and polychromed Methylene Blue	Jenner's method of preparing solution in methanol. Water added before applying stain	—	—
1901 Leishman (in England) (Baker, 1963, p. 267)	See above	See above; polychroming as in Unna's	—	—

	Composition	Notes	Use	Results
1901 Willebrand (Lillie, 1977, p. 495)	Sat. aq. Methylene Blue, 25 ml. 0.5% Eosin Y in 70% alc. 25 ml., 1% acetic acid 10–15 drops	First to use pH adjustments	Blood films	Nuclei dark blue, basiphil granules dark blue, erythrocytes red, eosinophil leucocyte granules red
1902 May-Grünwald	German equivalent of Jenner's	—	—	
1902 Wright	Same as Leishman	Polychroming different: heat mixture of Methylene blue and potassium bichromate for an hour in flowing steam	—	
1902 Giemsa (original)	$0.8^{\circ}/_{\circ\circ}$ aq. sol. Azure II 10 ml, $0.5^{\circ}/_{\circ\circ}$ aq. sol. Eosin Y 1 ml (recipe see Lillie, 1979, p. 497)	Azure II : Azure B (=I) + Methylene Blue (1 : 1); strengths of solutions not known	Malaria parasites in blood films	Neutrophil granules violet
1905 Hastings (Lillie, 1977, p. 496)	Polychromed Methylene Blue neutralised with acetic acid; *un*polychromed Meth. Blue; Eosin Y	Unpolychromed Meth. Blue added before eosinating	—	
1922 MacNeal (Lillie, 1977, p. 498)	In methanol: Methylene Blue chloride 1.0 g; Methylene Azure 0.6; Methylene Violet (Bernthsen) 0.2 g; Eosin Y 1.0 g	—	—	
1935 Kingsley (Lillie, 1977)	Same as MacNeal but different solvent: glycerol, ethanol	—	—	
1944 Lillie (Lillie, 1943b)	Half-converted Methylene Blue with calculated amount of $K_2Cr_2O_7$	Heating with $K_2Cr_2O_7$, in acid solution	—	
1979 Gilliland et al. (original)	Equivalent of Romanowsky	As Romanowsky; mixed with buffer pH 3.0 on slide to obtain pH 7.0	Blood films	Consistent results

Methylene Blue and the acid dye Eosin Y for this purpose. This combination was not new in itself, and was earlier described by Chenzinsky in 1888 as a blood stain. Both Romanowsky and Malachowski reported individually on their success in staining the nucleus of the malaria parasite differentially purple in blue cytoplasm in smears made from blood. For some reason Malachowski disappeared from the staining scene, but Romanowsky persisted. His name is still well known in the history of staining, and all the staining methods in which one or more thiazine dyes are combined with Eosin Y are called Romanowsky-type stains. They produce a differential staining effect: purple nuclei in blue cytoplasm. Whether Romanowsky really deserves all the credit given to him is doubted by some authors (Lillie, 1978).

The way in which Romanowsky and Malachowski prepared their staining solutions was different. Romanowsky used a mouldy, two-month-old Methylene Blue solution and mixed it with Eosin Y. Malachowski mixed his freshly prepared Methylene Blue solution first with alkali before adding Eosin Y. When he did not add alkali, he did not obtain the differential staining effect. Why it was so important as to how the stain was prepared or what actually happened to the dye over a period of time was not answered by the Russian scientists. However, it intrigued many other authors. In the following years, many attempts were made to produce a Romanowsky-type stain which would give consistent results. Since it was frequently used in tropical climates in the diagnosis of malaria in blood smears, it was very important that the stain should have a reasonably long shelf-life in hot climates. In the context of this it is important to realise that the majority of these stains were developed before the existence of refrigerators.

9.2.1 Modifications of the Romanowsky Staining Method

In contrast to the modifications of the Papanicolaou stain, each modification of the Romanowsky-type stains carries the name of its developer. From the varied list of recipes (table 9.1) we might

get the false impression that these staining methods differ fundamentally. Basically, they vary little. In all methods the cell smears are fixed by air drying, and post-fixed in methanol. In some methods, such as the May-Grünwald–Giemsa method, methanol is the solvent of the stock dye solution in which the smears are first stained. In order to qualify, the stain should produce the Romanowsky effect: purple-stained nuclei and blue cytoplasm. In addition, erythrocytes should stain pink and nucleoli blue. All other colours are the result of a combination of blue, purple and pink. However, the shades of colours vary between the different methods. Wright's stain, for instance, contains less Azure B, and results in a somewhat pale bluish-violet colour of the nuclei. The May-Grünwald–Giemsa method is mainly superior in haematology because of its subtle colouring of the cytoplasm of the various blood cells and their precursors (Lopez Cardozo, 1977).

Whether the Romanowsky-type stains are truly neutral stains in the sense that they involve a threefold action of coloured cation, anion and salt remains a matter of contention (see also section 9.3.1). We have included them in the section 'neutral stains' for historical reasons. In recent literature the most discussed variant is the Giemsa stain, especially because of its use in banding of chromosomes. The name 'Romanowsky–Giemsa stain' pays credit to both Romanowsky and Giemsa, and is used by the International Committee for Standardization in Haematology.

9.3 Romanowsky–Giemsa Stains in Current Use

Although the Romanowsky–Giemsa stains are widely used in haematology and cytology, few users will be concerned with the chemical background of the stains since they are commercially available. However, to understand the variable staining results, it is very important to know the chemical components of stains and the way they react with those of the cell. In addition, one should know what might happen to the stain over a longer period of time (see section 9.1 and the discussion of the work of Romanowsky and Malachowski). Most of the acquired knowledge is derived from the study of air-dried blood or bone-marrow smears. These cells are in a monolayer. Relatively little basic research has been done on the staining patterns in cells other than blood cells, although the staining method is widely used in other fields of cytology, especially in aspiration cytology (see section 10.2.5). The investigations concerning the differential staining of DNA and/or chromatin with these stains have intensified since the 1970s when the method became popular for the staining of chromosomes (G-banding: Sumner and Evans, 1973; Comings, 1975; Comings and Avelino, 1975; Sumner, 1980).

9.3.1 The Dyes

In their investigations of the Romanowsky method, Giemsa and his colleagues came to the conclusion that, after some period of time, the stain contained not only Methylene Blue and Eosin Y, but also several oxidation products of Methylene Blue together forming the group of thiazine dyes. Giemsa contended that the Azure B was responsible for the purple staining of the nuclei. Although it was, at that time, very difficult to produce completely pure Azure B, he tried to stain nuclei with as pure as possible a staining solution, and with this the nuclei (and hence the chromatin) stained purple. He concluded from this that it was not Methylene Blue but its oxidation product Azure B that caused the differential staining of the nuclei. Giemsa thus found why it was not possible for Malachowski to stain nuclei purple with a *fresh* Methylene Blue solution, and why Romanowsky's mouldy, two-month-old solution (containing oxidation products) worked (see section 9.2). Giemsa was certainly on the right track with his theories. Later spectrophotometric measurements have confirmed the presence of a number of Methylene Blue oxidation products in Romanowsky–Giemsa stains (Comings, 1975; Gilliland *et al.*, 1979). The differential staining of chromatin was confirmed in later years by other authors working with purified Azure B (Wittekind *et al.*, 1976, 1982; Wittekind, 1983). Chromosome banding can be

achieved by means of Azure B, and also with the other azures (Comings, 1975).

9.3.1.1 Thiazine Dyes

In the preparation of the Romanowsky-type stains, some thiazine dyes actually form part of the stain and others may be formed subsequently. In figure 9.1 the thiazine dyes are listed, together with their absorption maxima. These dyes vary in their number of methyl groups, which determines the place of the peak of its absorbance curve and thus its colour. Methylene Blue has its peak at 650 nm, and at the other extreme Thionin peaks at 598 nm. Thus staining solutions containing different proportions of the thiazine dyes will result in different staining patterns. Therefore, in order to achieve predictable staining results, the further oxidation of Methylene Blue and its derivatives needs to be kept under control (see section 9.3.2.2).

The staining solution used by Giemsa contained a combination of the thiazine dyes Methylene Blue and Azure B. He added Eosin Y, resulting in formation of Methylene Blue–Eosinate and Azure B–Eosinate. Lillie (1943b) combined Azure A–Eosinate, Azure B–Eosinate and Methylene Blue–Eosinate. He was hoping to make a stain of known quantities of active products of predictable qualities. However, even if the dyes are carefully measured, one still does not have a stable stain because the oxidation process goes on (see section 9.3.2). Even when fresh, the commercial Romanowsky–Giemsa stains contain all kinds of impurities (Marshall and Lewis, 1974). The standardised Romanowsky–Giemsa stain is discussed in section 9.3.4.

It is not only the presence of the different thiazine dyes that is of importance, but also their concentration in the staining solution. Comings (1975) was able to measure an increase in metachromatic properties of all the thiazine dyes involved in Romanowsky–Giemsa staining with increasing dye concentration, which means a shift from blue to red (see also section 3.9.3). In dilute solutions the monomer form prevails, and in concentrated solutions the dimer form. The aggregation of monomer dye molecules and dimer dye molecules (which

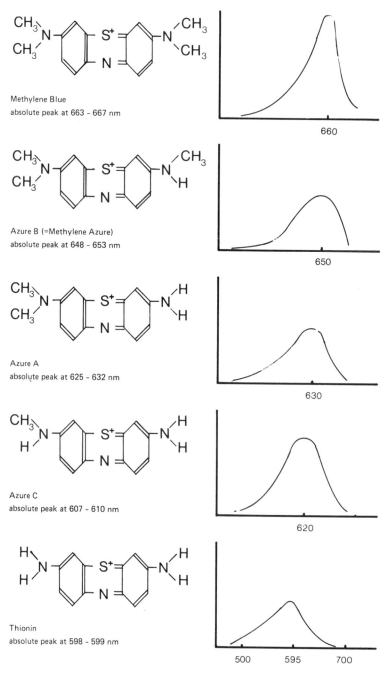

Methylene Blue
absolute peak at 663 - 667 nm

Azure B (=Methylene Azure)
absolute peak at 648 - 653 nm

Azure A
absolute peak at 625 - 632 nm

Azure C
absolute peak at 607 - 610 nm

Thionin
absolute peak at 598 - 599 nm

Figure 9.1   The oxidation sequence from Methylene Blue to Thionin and the subsequent shift in absorption maximum (after Comings, 1975).

happens in the nucleus and not in the cytoplasm of the cell) is the cause of the differential staining by the thiazine dyes. This is extensively studied by Galbraith *et al.* (1980) and Marshall *et al.* (1981), and further discussed in section 9.3.4.

9.3.2 Stabilising the Staining Solution

To obtain predictable staining results with the Romanowsky–Giemsa methods it is necessary to stabilise the staining solution. This means preventing the dyes from precipitating and stopping the oxidation process.

9.3.2.1 The Problem of Precipitation

When Methylene Blue and Eosin Y are dissolved together in water in the quantities first given by Romanowsky, the Methylene Blue–Eosinate that is formed precipitates. The precipitate can be dissolved by adding excess of either Methylene Blue or Eosin Y. Jenner (see table 9.1) dissolved the precipitate in methanol, to which Giemsa added glycerol for the same reason (Baker, 1963). Another way of solving the precipitation problem is by the addition of pure acid. Gilliland *et al.* (1979) found that at pH values lower than 4.0, precipitation did not occur at all temperatures, but at tropical temperatures (around 40°C), between pH 4.0 and 6.0, a precipitate is formed.

9.3.2.2 The Problem of Degradation of the Thiazine by Oxidation

Lowering the pH of the staining solution prevents not only precipitation but also degradation of the thiazine dyes by further oxidation (Gilliland *et al.*, 1979). Storing the stain solution at low temperatures is another way of slowing down the oxidation process (Dean, 1977). This is supported by Gilliland *et al.'s* studies, which were focused on keeping the stain stable in tropical climates and showed that there was a clear shift in the thiazine absorption peak when the staining solutions were kept at extremely high temperatures (40–50°C). The oxidation process is therefore favoured by high temperatures.

Liao *et al.* (1981) recommend the addition of diethylamine hydrochloride and dimethylamine hydrochloride to prevent oxidation. They claim a highly significant improvement of the shelf-life of such a methanolic Wright stain, a fact which is not confirmed by others in paper chromatographic studies (D.H. Wittekind, personal communication).

From table 9.1 it can be seen that others have attempted controlled oxidation by adding calculated amounts of acid or alkali. All methods that change the pH of the staining solution to prolong shelf-life require adaptation of the working solution to a pH around 7.0.

9.3.2.3 Pure Azure B: the Composition of the Standardised Stain

As discussed earlier, it is not the thiazine dye Methylene Blue that produces purple nuclei in blue cytoplasm but its oxidation products. Since it is so difficult to control the oxidation process of Methylene Blue, it seems much more logical to use the more stable Azure B, since this is the component responsible for the optimal differential staining effect of the Romanowsky–Giemsa stains. However, the problem is *how* to produce pure Azure B. Attempts to purify Azure B have been made by several authors (Löhr *et al.*, 1975; Marshall and Lewis, 1976) and it is now commercially available. Using this pure Azure B and Eosin Y, the International Committee for Standardisation in Haematology (ICSH) now recommends the following composition of the standardised Romanowsky–Giemsa stain: 3 g Azure B and 1 g Eosin Y are dissolved in 1 litre of methanol/dimethylsulphoxide (6 : 4). This stock solution keeps well and is diluted with a buffer of pH 6.8. A ratio of 1 : 15 is used as the working solution. The actual staining method is described in appendix 3.

9.3.3 The Romanowsky Effect: Purple Nuclei in Blue Cytoplasm

In the 1970s many investigators tried to unravel the mechanisms by which chromatin is stained purple. These investigations increased when the G-banding techniques for the identification of chromosomes became popular. In this method, by treating the chromosomes with salt and heat, specific areas are stained with the thiazine dyes whereas others take up no or almost no stain, resulting in 'banding' of the chromosomes. These banding patterns are characteristic for each type of chromosome. The images of thus-treated chromosomes stained with thiazine dyes can be compared with those stained with the Feulgen stain, which stains primarily DNA. In Feulgen-stained chromosomes, however, little or no banding is seen (Comings *et al.*, 1975), thus staining of the DNA with thiazine dyes must involve more than its mere presence.

9.3.3.1 Thiazine Eosinate

Sumner and Evans (1973) and Sumner (1980) contend that formation of a thiazine–Eosin compound and its precipitation play an important role in the purple colouring of the chromatin by the Romanowsky-type stains. If this compound is a salt, then the Romanowsky–Giemsa stains are correctly classified as 'neutral stains'.

According to Sumner and Evans' theories, the thiazine–Eosin precipitate is formed in those regions of the chromosomes where the DNA phosphate groups are at a specific distance from each other in such a way that they can react with two thiazine molecules. These two in turn can react with one Eosin molecule, which has two negative binding sites. Thus on these parts of the chromosomes which show purple bands when stained with the Romanowsky-type stains, a 2 : 1 thiazine–Eosin compound is formed. This theory is in keeping with Gilliland's ideas on thiazine–Eosin precipitate formation by side stacking along the DNA molecules (Gilliland *et al.*, 1979).

If it is true that the positive charge on the thiazine reacts with one of the negative charges on the Eosin molecule, thus forming the salt thiazine–Eosinate, then there is no positive charge left for its consequent binding (forming a salt) with the DNA phosphate anion. Therefore Sumner suggests that hydrophobic forces play some part in the links between DNA phosphate and thiazine. He bases this on the fact that the stain is preferentially found in hydrophobic regions of the chromosomes.

However, it is also possible to produce banding in chromosomes with thiazine dyes *only*, that is, in the absence of Eosin (Comings and Avelino, 1975; Comings, 1978). The banded staining pattern

(bands) of chromosomes is the same with or without Eosin. However, the resulting colour is not. In the presence of Eosin, the purple staining of the nuclei is enhanced and the colour further shifted to red. Azure B alone stains the nuclei more violet than purple.

9.3.3.2 Polymerisation of Thiazine Dyes: Causes of Metachromasia

When only thiazine is used, the colour is exclusively produced by the binding of the dimer form of thiazine. The degree of polymerisation of the thiazine dyes is higher in nuclei than in cytoplasm, probably due to polymerisation induction by DNA by side stacking of the dye along the DNA molecules. This causes the DNA to be coloured metachromatically. The stacking of dye molecules in the presence of DNA has the same effect as increasing the dye concentration in a solution, that is, metachromasia (see section 3.9.3). However, proteins like gelatin have the same polymerisation effect. The fact that proteins can also change the monomer form of the thiazine molecule to a dimer form permits the suggestion that, in addition, the nuclear proteins might be involved in the purple metachromatic staining of chromatin, which is a combination of DNA and nuclear proteins. The work of Comings and Avelino (1975) points to some action of the nuclear proteins in the binding of thiazine to DNA.

9.3.3.3 Effect of Fixation: the Role of Proteins

The Romanowsky–Giemsa method is predominantly applied on air-dried cells post-fixed in methanol. Comings and Avelino (1975) showed that fixation with methanol removes some of the histone and non-histone proteins from chromatin (see chapter 4). When these proteins are removed, more phosphate groups can react with the dye, which enhances the staining of chromatin by thiazine dyes. Furthermore, Comings and Avelino suggest that, in addition, salt–heat treatment used for the banding of chromosomes specifically alters the proteins in the 'band' areas in such a way that they cover the DNA more effectively, resulting in a diminished staining in the specific areas. As stated earlier, when the same chromosomes with the same 'banding' pre-treatment are stained with Feulgen, these variations in DNA staining in one chromosome are almost absent. Thus it seems likely that proteins play some role in the purple staining of the nuclei by thiazine dyes, but more research is needed to unravel the precise mechanisms involved.

Post-fixation in methanol of air-dried cells is essential for optimal differential staining of the nuclei and it should last for at least 15 min. If the slides are stained for a prolonged period of time (over 15 min) in the Giemsa staining bath, post-fixation continues, with concomitant changes in the chromatin patterns.

The Romanowsky–Giemsa method can also be applied on directly wet-fixed cells, as in the Papanicolaou method. The Romanowsky effect is then also visible; however, the staining pattern of the cytoplasm is slightly different. The cells and the nuclei have a more 'crisp' appearance, as in the Papanicolaou method, and the chromatin pattern somewhat resembles fixed Haematoxylin-stained nuclei.

9.3.3.4 Staining of RNA

One may wonder whether RNA is also stained metachromatically purple by the Romanowsky–Giemsa stain. Under certain conditions (Bennion *et al.*, 1975), RNA can be stained purple with azure dyes. However, Romanowsky–Giemsa if properly prepared and applied, stains nucleoli and RNA-rich cytoplasm of plasma cells orthochromatically blue. Thus, under these conditions, RNA is not stained metachromatically. (See Atlas, plate 12.)

9.3.3.5 Staining of Cytoplasm

The Romanowsky–Giemsa stain contains both an anionic (Eosin Y) and a cationic (thiazine) dye. As we know, the cytoplasm consists mainly of proteins, but in addition it may contain RNA. The orange-red colour of the erythrocytes and the protein coat of asbestos fibres (see plate 11) forms one extreme of the colour scale obtainable by the Romanowsky–Giemsa method. Here the anionic dye Eosin is predominant; due to linking with basic proteins. At the other extreme lies the blue colour of

the RNA-rich cytoplasm of plasma cells. Here the colour is due to the cationic thiazine dye, which has formed a link with RNA and, in addition, with acid proteins.

9.3.3.6 Summing Up

It is still not possible to tell exactly which events create the Romanowsky effect in cells stained with Romanowsky–Giemsa stains. The formation of a thiazine–Eosin compound in certain areas of DNA depending on a specific distance between the phosphate group on the DNA molecule is one possibility. Metachromatic staining of nuclear compounds seems obvious. This metachromatic effect is induced by DNA but can also be induced by proteins. It is very likely that all factors mentioned play their part in the mechanisms of the Romanowsky–Giemsa stains. In addition, others might play a role, such as competition of the dyes in penetrating the cytoplasm or nucleoli.

9.3.4 Staining Results of the Standardised Romanowsky–Giemsa Stain

Galbraith *et al.* (1980) and Marshall *et al.* (1981) used the standardised Romanowsky stain (section 9.3.1.1) for their elegant studies on the results of staining blood cells. They showed that in the stained cell components the dimer and the monomer of both Azure B and Eosin Y are present. The monomer of Eosin has an absorbance peak at wavelength 515 nm and the dimer a less distinctive peak at wavelength 479 nm (both red). The peaks of the monomer and dimer of Azure B lie at respectively 596 and 513 nm. By analysing the absorbance spectra of stained blood cells, Galbraith *et al.* came to the conclusion that all four components can be found in the cells but their ratios differ. The end result of the staining is caused by a summation of the staining effect by the four components individually. The authors do not mention the formation of the salt Azure B–Eosinate. They found that polymerisation in nuclei is higher than in cytoplasm: this holds for Eosin Y as well as for Azure B. The cytoplasm, if it has a negative charge, binds with the positively charged monomer form of

Table 9.2   Staining effect of the standardised Romanowsky–Giemsa stain on blood films

Acid components	Colour
Basiphil cytoplasm	
polymorphonuclear leucocytes	blue
lymphocytes	blue
monocytes	blue
plasma cells	blue
parasitic protozoa	blue
Chromatin	
polymorphonuclear leucocytes	purple
lymphocytes	purple
monocytes	purple
plasma cells	purple
parasitic protozoa	red
Acidophil cytoplasm of	
polymorphonuclear leucocytes	pink
Acidophil granules	pink
Red blood corpuscle	pink
Neutrophil granules of cytoplasm	
of polymorphonuclear leucocytes	purple

Table 9.3   Staining effect of the standardised Romanowsky–Giemsa stain on other than blood films

Material	Colour
Keratinised cytoplasm	azure blue
Mucin	reddish-blue
Collagen	purple–reddish
Trichomonad cytoplasm	blue
Trichomonad nucleus	purple
Monilia	dark blue
Chlamydia	red
Striated muscle	bright blue
Myxomatous background	pink
Cilia	purple

Azure B. The erythrocytes and the granules in leucocytes, which are positively charged, are stained by the negatively charged Eosin. They also showed that, in the presence of Eosin, the uptake of Azure B dimer (purple) is enhanced, as also is the uptake of Azure B monomer (blue) although to a lesser extent. (See tables 9.2 and 9.3).

Galbraith *et al.* have not investigated the staining of nucleoli. These are stained in all shades of blue in the standardised stain, ranging from dark blue to azure. In dense and darkly stained nuclei they are not clearly visible (see Atlas, plates 5 and 12).

9.3.5   Staining Times

The Romanowsky effect needs some time to develop. If the cells are stained for too short a time, the nuclei remain blue. It is virtually impossible to 'overstain' the nuclei, as can be the case in the Haematoxylin method. Depending on the material, the staining times with the standardised Giemsa stain vary from 15 to 45 min. If the results are not satisfactory, the smear can be restained.

9.3.6   Influence of pH on the Staining Solution

The Romanowsky effect occurs roughly between pH 6.0 and 8.5 of the staining solution. At low pH (6.0–6.5) the colour of the erythrocytes is bright red. The staining of the erythrocytes becomes less at higher pH, and at pH 7.5 they are colourless. With further increase in pH they become bluish green. At pH between 6.0 and 5.0 everything except erythrocytes stains blue, and the Romanowsky effect decreases rapidly. At pH lower than 5, everything stains red with the Eosin. These effects are true for thinly spread cells. The staining effects in thick parts are unpredictable, and always (between pH 5.5 and 8.5) towards the blue side of the colour spectrum. (See Atlas, plate 24.4.)

9.3.7   General Remarks on the Staining Results of the Romanowsky–Giemsa Method

All these investigations on the results of the

Romanowsky–Giemsa stains mentioned above were done on blood cells and chromosomes. Little is known about the role of density of the cytoplasm in differential staining, as has been widely discussed for the Papanicolaou method. It is very possible that some competition mechanism in penetrating the cytoplasm, which is so important in the Papanicolaou method, plays an additional role in the staining results.

The Romanowsky–Giemsa staining method should be used for thinly spread, air-dried cells. Often some parts of the smear will be too thick: in these the purple differential staining of the nuclei is absent, and everything stains blue. The thinner parts of the material are often found in the outer zones of the smear. Parts which are too thick, or thick cell groupings in which the nuclei are blue, can be destained in methanol, and consequently restained with a higher concentration of the dye, and/ or stained for a longer period. (See Atlas, plate 9.)

Air-drying does not, unlike wet fixation, result in a clear separation between euchromatin and heterochromatin (see chapter 4). Some separation of hetero- and euchromatin occurs in post-fixation, but a different nuclear chromatin pattern is created. Thus the air-dried Romanowsky–Giemsa stained nuclei are less transparent compared with wet-fixed Papanicolaou stained nuclei, and there is no condensation of chromatin on the nuclear envelope. Post-fixation with methanol is necessary to enhance the nuclear staining pattern: if the post-fixation is too short ($<$ 2 min) the nuclear details of malignant cells are not visualised. If the preparations are stained over a long period of time, the nuclei become more transparent and the difference between heterochromatin and euchromatin is more clear. When fresh cells come into contact with formalin vapour, everything stains blue, including erythrocytes. In small quantities it enhances the blue staining of the nucleoli.

Each modification has its different shade of colours, as visualised in the Atlas. In addition, since slides made from material other than blood and bone marrow are very uneven, the staining results in one slide might vary greatly between different microscopic fields.

Many questions remain unanswered: why do components other than chromatin (i.e. neutrophil granules in polymorphs and cilia) stain purple? Why does the cytoplasm of keratinized cells and striated muscle stain azure blue and not dark blue? Why do erythrocytes stain green at high pHs?

Preparing Cytological Slides

10.1 Introduction

This chapter is the 'cookery book' part of this volume. First, it will give practical information on how to process cytological material from various sites; how to preserve and fix cells; how to make cytological slides; how to process difficult-to-handle material; how to combine the various fixing and staining methods; and how to prepare material for adjuvant scanning and transmission electron microscopy.

We have tried to make this chapter as practical as possible but we realise that it is not complete. There are various other modifications of the techniques, which we have not listed: the choice in the literature was abundant and therefore we have to make an admittedly highly personal choice.

10.2 Processing Cytological Materials

10.2.1 Gynaecological Material

10.2.1.1 Cervix

The Ayre spatula or one of its myriad modifications is most commonly used to prepare cervical smears (see section 10.3.1.1). Self-smear methods have been developed, using either irrigation (Davis, 1962) or a sponge to be processed in the laboratory (Arata et al., 1978).

10.2.1.2 Endocervix

The endocervix can be either aspirated or brushed. We use the brush (Cytobrush) developed by Stormby in Sweden. It has proved in our hands to be highly successful on an outpatient basis. The material collected on the brush can be smeared or rinsed (section 10.2.4).

10.2.1.3 Endometrium

Special sampling devices are developed for the endometrium using irrigation, aspiration or brushing. The material from the endometrium can also be collected by using cellular material taken from an IUD or collected on a vaginal tampon, or by aspirating the endocervix. The devices for sampling endometrial cells can be summarised as follows:

(1) Vakutage device (Warner-Chilcott, Morris Plains, New Jersey)
(2) Isaacs Cell Sampler (Kendall Co., Chicago, Ill.; Isaacs and Wilhoite, 1974)
(3) Gravlee Jet Irrigator (Upjohn Company, Kalamazoo, Mich.; Gravlee, 1969)
(4) Medhosa Cannula (Etalissements Carrieri, Paris, France; Jimenez-Ayala et al., 1975)
(5) Vaginal tampon (Couture et al., 1979)
(6) Mi-Mark Helix (Simpon-Bayse Inc., Wilmington, Delaware)
(7) French Balayette Brush (Laboratoire CCD, Paris, France)

Table 10.1 Evaluation of endometrial sample techniques (Schachter *et al.*, 1980)

	No.	*Technical failures (%)*	*Specimens collected*	*Satisfactory*
Mi-Mark Helix	32	12.5	28	25
Gravlee Jet Wash	52	8.3	48	44
Lippes Loop IUD	40	2.5	39	35
French Balayette Brush	56	1.8	55	54
Total	180	5.6	180	158

(8) Endocervical aspirator
(9) Lippes Loop IUD (Bercovici *et al.*, 1958)

The different techniques differ mainly in the percentages of technical failures and amount of material collected (table 10.1). The endometrial material can be suspended (section 10.9), cell-blocked (section 10.5) or smeared (section 10.3.1).

10.2.2 Urine

Freshly voided urine is the optimal material for urinary cytology (Beyer-Boon, 1977). If it is not possible to process the urine immediately, the following measures can be taken:

(1) Add as a preservative a bacteriostatic agent, for instance thiomersal (Beyer-Boon *et al.*, 1979b).
(2) Add a preservative fluid to the specimen with or without pre-fixing action (table 10.2).

Although degeneration is delayed by ethyl alcohol, it is not fully stopped. Crabtree and Murphy (1980) even contend that it is not advisable to add 50% ethyl alcohol. Higher grades harden the cells too much, making cytopreparation difficult. Beyer-Boon *et al.* (1979b) contend that in the first three days cellular degeneration can be attributed to the deleterious effect of the toxins produced by the multiplying bacteria in the urine. The good results obtained after using preservatives such as van der Griendt's solution can probably be explained by

some additional bacteriostatic action that they possess.

Table 10.2 Preservatives and their applications

Preservative	*Application*
van der Griendt's fluid	Urine, sputum, aspirates
Esposti's fluid	Urine
Saccomano's fluid	Sputum, aspirates
Delft suspension medium*†	Aspirates, CSF
Leiden fixative	Aspirates
Davis's irrigation fluid	Endometrial irrigation
Mucolexx	Mucoid material, aspirates
Ethyl alcohol 50%	Urine added to bronchial lavage with physiological saline
RPMI Medium 1640*	Aspirates
Marsan's fluid*	Bronchial brushings
Merthiolate† (thiomersal)	Urine

* Does *not* contain fixative!
† Is, or contains, a bacteriostaticum.

10.2.3 Sputum

10.2.3.1 Pick-and-smear Method

The fresh or pre-fixed sputum is put in a petri dish. With tweezers, areas of the specimen that are most likely to contain cancer cells (bloody brown or brown–yellow streaks) are picked up. Smears are made from these, preferably using the sandwich technique (see section 10.3.1.2).

10.2.3.2 Saccomano's Technique

The sputum is collected in Saccomano's fluid (Saccomano *et al.*, 1963) in a tube. The contents of the tube, increased if necessary by 50% ethyl alcohol, are transferred to a blender and blended for 3–4 s to liquefy the mucus to a cloudy, liquid, homogeneous state. If this state is not reached after 3–4 s, the material should be blended for an additional 2–3 s. It is then transferred to another 50 ml centrifuge tube and centrifuged for 15 min at 1500 r.p.m. The smeared sediment should be post-fixed with 95% ethyl alcohol or alcohol–polyethylene glycol.

10.2.3.3 Cell Blocks

In general, staining times for paraffin-embedded-section cells are much longer than for smeared cells. For the Papanicolaou methods, the staining times should be multiplied by 2, and for the Romanowsky–Giemsa method by 4. These long staining times can be shortened considerably by staining in staining dishes at 35°C. One should experiment to obtain the desired results.

The sputum is fixed in either neutral formalin or in a solution of 13 g picric acid dissolved in 70% ethyl alcohol. Formalin fixation is shortened from 24 h to 5 min if the plastic container with formalin and sputum is set in the microwave oven at high-energy level. The block is cut at three levels.

10.2.3.4 Comparing the Three Methods

In the pick-and-smear method, the smears contain strands of mucus in which the cancer cells are caught. These are broken down in the Saccomano method. However, in the latter there is very little overlap of cells, and the cells are evenly divided over the slides. In the pick-and-smear method, the technical skill of the person who picks up the parts of the sputum to be processed is decisive, whereas in the Saccomano method there is no selection. These two methods were compared by McLarty *et al.* (1980): in the pick-and-smear technique, a higher percentage of cases with ferrugenous bodies was found, and the Saccomano technique was more successful in the detection of atypical cells (table 10.3).

In the cell block technique, as in the pick-and-smear technique, the strands containing cancer cells are left intact. The advantages of the former method are discussed in section 2.6.1. For a comparison of the smear and cell block techniques, the reader is advised to consult the study of Gray (1964). In general, a tissue pathologist will be able to reach a higher level of refinement of diagnosis with the cell block method, and a cytologist trained in cervical cytology will obtain better results from the smear technique. Both methods, when well performed, result in excellent morphology.

Table 10.3   Comparison of Saccomano's technique and the pick-and-smear technique (McLarty *et al.*, 1980)

	*Abnormal cytology**	*Ferruginous bodies*
Pick-and-smear	2.0%	18.3%
Saccomano	4.9%	12.9%

* $P < 0.0001$.

10.2.3.5 Van der Griendt's Method

Van der Griendt's method for sputum is as follows: the sputum sample is suspended in van der Griendt's fluid (see table 10.2) and processed by means of the sedimentation technique (see section 10.4.3 and figure 10.7). Another modification is the use of 70% alcohol and polyethylene glycol as suspension fluid. Shake the container well when the fresh sputum is brought into the suspension fluid; the mucoid material will thus be broken up into very small particles that are fixed sooner and are more easily processed.

10.2.4 Brush Cytology

With the inroduction of transfibre-optic equipment in the inspection of the bronchial tree (Bibbo *et al.*, 1973) and the digestive system, the ability to diagnose cancer of the lung and of the digestive system by brushing the visualised lesion and processing the collected cytological material has improved remarkably. The Japanese investigators in particular have perfected these methods (Endo *et al.*, 1974). There are two main methods for processing the material collected on the brush.

10.2.4.1 Brush Smearing Technique

For details of the brush smearing technique, see section 10.3.1.5.

10.2.4.2 Brush Rinsing Technique

Insert the brush in either preservative solution (see table 10.2) or a balanced salt solution. When the material contains a lot of mucus, Mucolexx can be used as a rinsing solution (see section10.6.1). Smith *et al.* (1980) reported that, in lung cytology, when the brushes are rinsed after smearing, 26% more of the specimens were diagnostically satisfactory.

10.2.5 Aspiration Cytology

Aspiration cytology is known under many names (see section 2.5.1.1). The technique of performing an aspiration is visualised in figure 10.1. The negative pressure in the needle should be present while aspirating, during which time the needle should be moved up and down and in several directions in the lesion. After the cytological material has been collected in this way, the negative pressure in the needle can be released and the needle withdrawn. The majority of the aspirated material

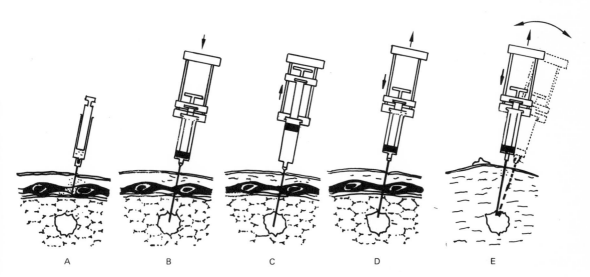

A   B   C   D   E

Figure 10.1 Aspirating lumps. (A–D): in the thorax. (A) anaesthesia; (B) needle enters lump; (C) producing negative pressure: material in needle; (D) releasing negative pressure; then the needle can be removed. (E) in the breast: when needling, the angle of the needle can be changed during the period with negative pressure. No local anaesthesia is needed.

is *in* the needle, and thus not visible to the aspirator during the aspiration. When a blood vessel is aspirated, blood is visible in the syringe: the negative pressure should then be released immediately because an admixture of aspirated cells with blood is not wanted. Thompson (1982) showed that the essence of fine needle aspiration is the combination of simultaneous negative pressure *in* the needle and the needle movements. Negative pressure applied with the needle stationary should be avoided, since it is likely to produce dilution with blood or other fluids (see figure 10.2).

As stated above, the main part of the aspirated material is *in* the needle. If the aspiration is not performed by the cytologist, and the cytology department is close by, the needle and its contents can be sent directly to the department so that the cytopreparatory techniques can be performed (figure 10.2). If this is not the case, the aspiration can be suspended (see section 10.9), or smears can be made (section 10.3.1.2).

10.2.6 Cerebrospinal Fluid

Cerebrospinal fluid (CSF) contains few cells and

has a low protein content. It should be processed with great care, and if it cannot be processed directly, a preservative should be added. For this purpose, we advise using an equal amount of Delft fluid (see table 10.2). The cell concentration technique should be a gentle one, if possible one of the sedimentation techniques (10.4.3) or the Cytospin method (section 10.4.2: Kölmel, 1977). It was only after the introduction of these two methods, in which final desiccation of the cells in a non-physiological method is circumvented, that the art of CSF cytology was able to develop fully (McCormick and Coleman, 1962; Krentz and Dyken, 1972).

10.2.7 Coelomic Fluids

In general the coelomic fluids, ascites and pleural and pericardial fluids, have a high protein content and therefore the cells do not degenerate quickly. The fluids can be kept in the refrigerator for a day or two. We prefer *not* to prefix the fluids. In coelomic fluids, fibrin clots are often formed. In these the cells are caught; we do not stop clotting because smears and blocks can be made from the

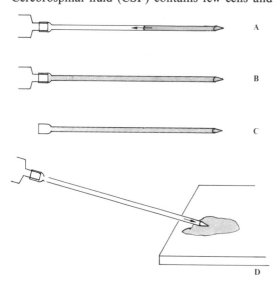

Figure 10.2   **Sample in the needle. With negative pressure (A, arrow) the needle is filled with cell material (B). The needle with the sample in it can be brought (in a horizontal position) to the laboratory (C) and there emptied for making smears (D).**

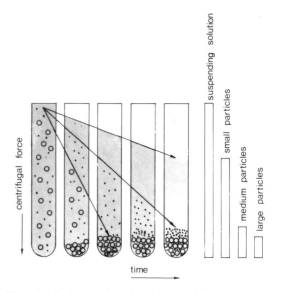

Figure 10.3   **Distribution of cells in centrifuge tube before, during and after centrifugation. Note that cancer cells (large particles) end up on the bottom.**

clots. However, others prefer to remove the clots from the fluid. The protein content of serous fluids is frequently far higher than that of blood and can be well over 20 g per 100 ml. The increased viscosity of the fluid then interferes with the flattening of the cells, and the staining is changed, not only that of the background but also that of the cells themselves (To *et al.*, 1983). Therefore it is advisable to wash the cell pellet first with PBS if it has a high protein content, and then make cytological preparations. Another problem might be the admixture of blood, as discussed in section 10.6.2.

One should be aware that the cell population in the fluid might vary depending on where the sample is taken: above the diaphragm or in the upper levels of the fluid. In the first, papillary dense cell groupings predominate, and in the latter, large isolated lipid-filled malignant mesothelial cells (see figure 10.3).

10.2.7.1 Removal of Clots in Pleural Fluid

To remove the clots, the following procedure should be adopted.

(1) Add 10 ml of reconstituted Varidase Tepical (Lederle UK) to the specimen.
(2) Wait 5 min during which time the clots will loosen and the trapped cells will be freed.
(3) Remove the loosened clots after shaking the specimen.
(4) Wash the freed cells twice with PBS.

10.2.7.2 Preparing Cytological Slides from Coelomic Fluids

After washing the cell pellet (see section 10.2.7) and removing the blood (10.6), the pellet is resuspended in physiological saline or suspension fluid and the cells are concentrated using one of the methods discussed in section 10.4. If the specimen contains many polymorphs and lymphocytes, it is advisable to use a gradient technique (see section 10.4.4). From the clots, cell blocks or imprint smears are prepared.

10.2.8 Special Collection Techniques for the Digestive System

10.2.8.1 Balloon Techniques for the Diagnosis of Cancer

Balloon techniques for the diagnosis of cancer of the oesophagus and stomach are used mainly in China (Shu, 1983). A small inflatable rubber balloon attached to soft tubing is introduced. The balloon has an abrasive surface, which, after inflation, scrapes superficial layers from the mucosal lining. It is a simple method, to be used on an outpatient basis.

10.2.8.2 Endoscopic Irrigation and Aspiration for the Detection of Carcinoma of the Hepato-biliary Tract

Endoscopic irrigation of bile ducts can be performed and the cellular contents studied (Nishimura *et al.*, 1973). The pancreas can be aspirated using a duodenal fibroscope (Endo *et al.*, 1974).

10.3 Various Slide Preparation Techniques

10.3.1 Smear Preparation Techniques

10.3.1.1 Material Collected with a Wooden or Plastic Spatula

Most gynaecological smears are made with a wooden or plastic spatula. Many clinicians spread the cells by smearing the material collected on the spatula vertically (figure 10.4). A minority (around 5%) use the zig-zag method and very few clinicians (less than 1%) smear the material clockwise. Rubio *et al.* (1980) showed that the way the smear is made influences cell harvest on the slide. They contend that some of the factors influencing the accumulation of tumour cells in certain areas of the slide may be: unequal pressure of the spatula during smearing, variations in the angle between the instrument and the surface of the slide, and the displacement of tumour cells to certain areas by overlapping strokes. In a routine laboratory, the screener can often recognise which doctor has made the smears: most have an individual style of smearing.

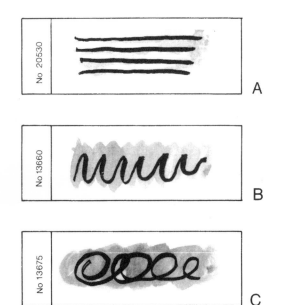

Figure 10.4   Three methods to make a smear: (A) smearing vertically, (B) zig-zag method, (C) clockwise.

Rubio also showed that many tumour cells are left on wooden spatulas (Rubio, 1977). It would therefore seem advisable to use plastic spatulas. However, in our experience patients bleed much easier when a plastic spatula is used, and the collected material glides easily from it; we therefore prefer wooden ones.

10.3.1.2 Material Collected by Aspiration

Aspirates contain unfixed cells, so care must be taken not to crush them. There are several ways to make a smear from an aspirate.

(1) The thick pieces are squashed gently with a second slide, and the remainder is distributed over the slide (figure 10.5). If some material is left, a second smear can be made with it.
(2) The material is sandwiched between two slides.

It is *always* necessary to smear the aspirate. In the case of wet fixation, every minute counts! Thus, the cap of the spray fixative container *must* be opened *prior* to aspirating. Material without mucus dries very fast: in these cases it is advisable to put some spray fixative on the glass slides prior to aspirating (see section 10.10).

The needle containing the sample can also be brought directly to the cytology laboratory (see figure 10.2).

10.3.1.3 Smears Made from Sediments

The same techniques can be used as are employed following collection by aspiration. The supernatant

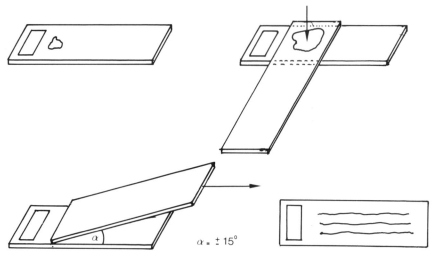

Figure 10.5   Preparing slides from an aspiration. First, the thick parts should be squashed with a second slide. The remainder of the material is distributed: note angle of the slide!

should be removed *completely* and only the cell pellet should be used for the smears. Take care to use the *whole* pellet: the cancer cells lie on the bottom of the tube (figure 10.3).

10.3.1.4 Smears Made from Sputum

For the pick-and-smear technique see section 10.2.3.1. The sandwich technique (10.3.1.2) is to be preferred for mucoid material such as sputum.

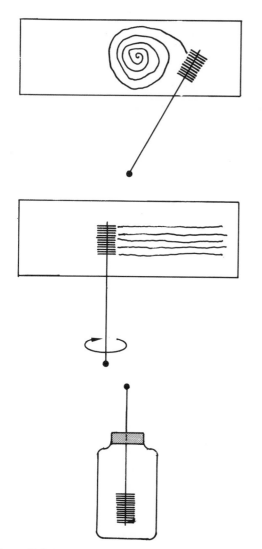

Figure 10.6 Preparing smears from material collected on brushes, or suspending it. For smearing, use circular or vertical method.

10.3.1.5 Smears Made from Material Collected on Brushes

There are two ways to make smears from material collected on a brush (see figure 10.6):

(1) Smear the brush on a glass slide using a circular motion over an area of about 1.5 cm diameter (for wet-fix smears).
(2) Roll the brush over the glass slide (for air-dried smears).

10.4 Cell Concentration Techniques

As described in section 2.5.3, cells can be concentrated from body fluids and cell suspensions in various ways. The cells can be in suspension in body fluids, CSF or urine, or may be suspended in a preservative fluid by rinsing the brush or thin needle (see section 10.9). The cells can be either unfixed or pre-fixed. For treatment of coelomic fluids, see section 10.2.7; of CSF, see section 10.2.6; and of urine, section 10.2.2. The concentrating methods are subdivided into five categories, which we will now consider.

10.4.1 Preparing a Cell Pellet in the Concentrational Centrifuge

The supernatant is completely removed and smears are prepared from the cell pellet (see section 10.3.1.3). For the complete removal of the supernatant, the centrifuge tubes can, after centrifugation, be placed upside down with their open ends on a blotter. The fluid will thus drip down but the cell pellet will stick to the centrifuge tube.

10.4.2 Concentrating Cells on the Slide

In the commercially available cytocentrifuges, Cytospin I and II, the fluid is absorbed in the filter paper during centrifugation, and cells become stuck to the glass slide. It is advisable to add polyethylene glycol–alcohol to the suspension, and to post-fix the slides with spray fixative after centrifugation. In Leif's method (Leif *et al.* 1977), the cells are first glued to the slide by centrifugation, and then the supernatant is removed. Here also it is advisable to add polyethylene glycol–alcohol to the suspension and to post-fix with spray fixative.

10.4.3 Sedimentation Techniques

Sayk's original method (Sayk, 1962) was based on sedimentation of the cells by force of gravity while the fluid was absorbed by filter paper. Later modifications by Bots *et al.* (1964) and van der Griendt (1984) are based on the same principle. It is a very gentle method: the cells fall like snowflakes on to the glass slide. Therefore it is used in refined cytodiagnosis (e.g. for processing CSF). However, the cell loss is considerable (Boon *et al.*, 1980). Many simple instruments have been designed based on Sayk's principle (Bots *et al.*, 1964; Beyer-Boon, 1977).

In van der Griendt's method (1984) he uses a brass ring placed on filter paper with a hole the same size as the inner circle of the ring (diameter 15 cm). The ring is filled with the cell suspension. He uses as a suspension medium van der Griendt's fluid, and thus urine, aspirates, body fluids and sputum (pressed through a tea strainer) can be processed. The small mucoid fragments left behind in the strainer can be smeared, or slides can be made with either the squashing technique (section 10.7.3 and figure 10.12) or with the imprint technique (section 10.7.1). The area on the slides with cells is larger than in the Cytospin method. Care should be taken to post-fix the cells, either in suspension or on the slide or both. (See figure 10.7.)

10.4.4 Gradient Techniques

In gradient techniques the cell suspension is added to a Percoll solution with different gradients. In our laboratory, we prepare a continuous gradient as in immunological methods (see figure 10.8) but a discontinuous gradient can also be used (Nagasawa and Nagasawa, 1983). The Percoll with the cell suspension on top is centrifuged: different cell types are concentrated in different levels. Each cell layer can be processed separately either by the smear

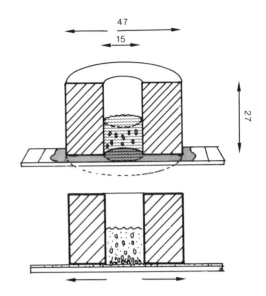

Figure 10.7   Sedimentation using van der Griendt's method. Brass rings (inner diameter 15 cm, outer diameter 47 cm) are placed on a filter paper with a hole of 15 cm diameter. The brass ring is filled with the cell suspension. The fluid is slowly absorbed in the filter paper (see arrows). For sputum, the material should be filtered through a tea strainer prior to sedimentation. The slides should be post-fixed.

technique, or by resuspension. These suspensions can be processed by the Cytospin or sedimentation techniques. Gradient centrifugation should be used if diagnostic cells are admixed with large numbers of non-diagnostic cells. This might be the case in pleural effusions (see section 10.2.7). The diagnostic cells should be concentrated to reach a diagnosis. This can be achieved by making use of the fact that the density of malignant cells differs from that of the non-diagnostic polymorphs, lymphocytes and erythrocytes. To reach the proper diagnosis, the cytologist has then only to screen a few slides of the enriched fraction. Figure 10.9 shows the cell population of an unfractioned pleural fluid and of the separate fractions (Nagasawa and Nagasawa, 1983).

10.4.5 Filter Techniques

Filter techniques to concentrate cells from body fluids have become very popular since the 1960s

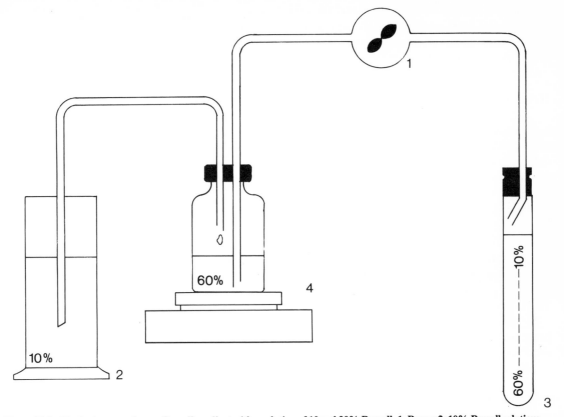

Figure 10.8   Producing a continuous Percoll gradient with a solution of 10 and 20% Percoll. 1: Pump; 2: 10% Percoll solution; 3: continuous Percoll gradient ranging from 10 to 60%; 4: 60% Percoll solution.

(Bernhardt *et al.*, 1961; Reynaud and King, 1967). The main techniques used are the Millipore and Nucleopore techniques.

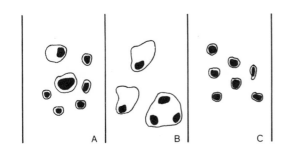

Figure 10.9   Smears from separated cell populations using gradient technique: (A) original mixed population; (B) cancer cell population; (C) lymphocyte fraction.

10.4.5.1   Millipore Filter Technique

Millipore filters consist of cellulose esters. They are manufactured by the Millipore Filter Corporation. There are two types: MF, with a pore size of $8.0 \pm 1.4\,\mu\text{m}$, and SM with a pore size of $5.0 \pm 1.2\,\mu\text{m}$. The cellulose matrix occupies 15–20% of the total filter. The following method is described by Koss (1979).

(1) Place the filter in the apparatus (see figure 10.10).
(2) Moisten the filter with physiological saline.
(3) Decant fluid in filter cup. If the fluid specimen is less than 20 ml dilute it with physiological saline.
(4) Filter fluid under negative pressure (20–50 mmHg) (using water pressure).

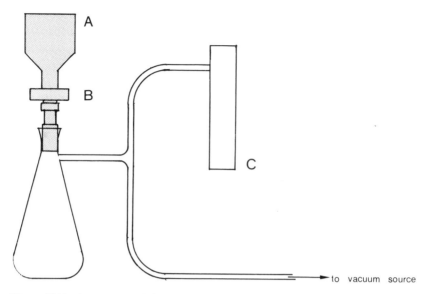

Figure 10.10   Apparatus used for Millipore filtration. A, filter cup; B, filter holder; C, manometer. The negative pressure is produced by a vacuum source.

(5) Filtration is stopped when there is still a small amount of fluid in the filter cup.

(6) Trim the filter membrane such that it fits on a glass slide, and attach it to the slide with a paperclip, or use special staining clamps.

(7) Fix the filter in 95% ethyl alcohol for 30–60 min.

(8) Stain the slide with the chosen Papanicolaou method; after the EA, put the filter in 100% propyl alcohol for 2 min, and then in 95% ethyl alcohol (100% ethyl alcohol softens the filters).

(9) Mount the filter with an adequate amount of mounting medium to prevent air bubbles, which would make the filter opaque.

10.4.5.2 Nucleopore Filter Technique

Nucleopore filters consist of a polycarbonate plastic film. They are manufactured by the General Electronic Company. The steps are the same as steps 1–7 of the Millipore filter technique. In step 8 the time in xylene should not be over 15 min, otherwise the filters start to curl. Nucleopore filters can also be fixed with air drying for the Romanowsky–Giemsa method. For that purpose the filters are attached to the glass slide with a paperclip, and,

when dry, post-fixed in methanol and consequently stained. Nucleopore filters can be dissolved in chloroform (see Koss, 1979).

10.4.5.3 Filter Imprint Technique

In contrast to smeared cells, the cells on filters are evenly distributed (Nielsen, 1972). This aspect is preserved when cell imprints are made from material collected on a Millipore filter (see figure 10.11). The filters are not stained and filed in this method.

The method according to Fischer (1978) is as follows:

(1) Coat slide with adhesive (see section 10.8).

(2) Filter the material through a rectangular Millipore filter with 8 µm pore size.

(3) Place the filter face down against an albuminised slide, applying an even pressure.

(4) Remove filter and let the slide air-dry or fix with coating fixative.

Boccato (1981) adopted the following procedure:

(1) Store slides in deep freezer.

(2) Filter material through Millipore filter (prefixed or not).

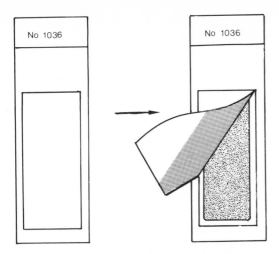

Figure 10.11 **Making an imprint slide from a filter. In this method it is not necessary to stain and mount the filter. Filing the slide without the filter is easier than with it.**

(3) Place the filter face down against the cold slide taken from the freezer (within 30 s).
(4) Press the slide evenly against the frozen slide by applying pressure, for instance with an ink roller.
(5) Remove the filter and fix immediately with 95% ethyl alcohol.

10.4.5.4 Special Measures for Coverslipping Millipore Filters

In Millipore filters air can be trapped. This makes them less transparent (see section 10.8). To prevent this from occurring, it is recommended that the following measures be taken:

(1) Pour mounting medium (Permount or Eukitt) in a shallow dish (e.g. a milk-testing dish).
(2) Let the stained filters soak in the mounting medium in the dish for a minimum of 5 min.
(3) Gently place the soaked filter on the slide, being careful to avoid trapping any air bubbles under the filter.
(4) Place a coverslip over the filter without using any extra mounting medium, and avoid trapping any air bubbles.
(5) Let the coverslipped filters 'cure' for a few hours prior to microscopical evaluation: this

allows the mounting medium to settle and the coverslip will be more firmly glued on the slide.
(6) Place the filters for 24 h in a 70°C oven before filing them. It is best to place the slides in a horizontal file for a month before placing them in a vertical file. If these precautions are not taken, the slides will stick together.

10.5 Cell Blocking Techniques

Clots in body fluids, sputum and endometrial material are well suited for the application of cell blocking techniques. The material is fixed in neutral formalin or picric acid–ethyl alcohol. For fixation of sputum see section 10.2.3. The fixed material is wrapped in a gauze or paper tissue and processed as histological material. Sections are made on three levels, and ribbons are kept for additional staining. The sections can be stained with Haemotoxylin and Eosin, and for sputum, with the Leiden–Orange G Papanicolaou method.

10.6 Liquefying Mucus and Haemolying Blood

Handling mucoid or very bloody material can have problems. In these cases it is advisable to liquefy the mucus or to haemolyse the blood.

10.6.1 Mucolysis

Cytological material can contain a lot of mucus, which interferes with optimal cytopreparatory technique. The mucus can be liquefied with acetyl-cysteine (Boccato, 1981) or with the commercially available Mucolexx. The latter contains in addition polyethylene glycol–methanol, and can therefore also be used as a preservative. Mucoid aspirates and brushed material can be suspended in it. In addition, the mucus can be liquefied by blending, as in the Saccomano technique (see section 10.2.3.2).

10.6.2 Haemolysing solutions

Erythrocytes, disturbing if present in the Papan-

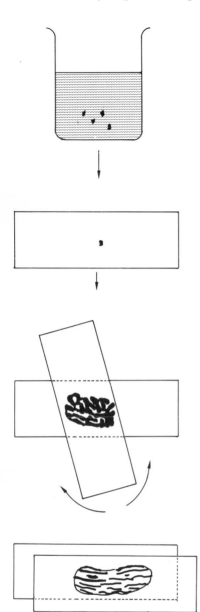

Figure 10.12 Squashing (or sandwich) technique, here shown for small pieces of histological material. The histological pieces are squashed until they consist of only a few cell layers. The sandwich technique can also be used directly for mucoid material such as sputum, or indirectly when van der Griendt's method for sedimentation is used (see figure 10.7) and mucoid material is left in the tea strainer.

icolaou staining method, can be haemolysed by means of the following solutions.

Acetic acid (Pool and Dunlop, 1934)
Saponin (Nedelkoff *et al.*, 1961)
Streptolysin O (Malmgren, 1965)
2 M urea (Pieslor *et al.*, 1979; can be applied on stained slides)
Formol saline 0.1% (van der Griendt, 1984; can be applied on Papanicolaou air-dried slides)

Saponin and streptolysin O are mainly used to haemolyse erythrocytes in Papanicolaou-stained slides. The procedure is then as follows:

(1) Remove the coverslip with xylene.
(2) Process through graded alcohols to water.
(3) Place the slide in 1 M urea (1 min to 24 h).
(4) Place the slide in 95% ethyl alcohol.
(5) Restain the slide.

Erythrocytes on air-dried slides can also be haemolysed. The air-dried slide is incubated for at least 2 h in van der Griendt's solution prior to staining by the Papanicolaou method. The great additional advantage of van der Griendt's method, in which the chromatin pattern is changed, is that all erythrocytes thus disappear, making screening of Papanicolaou slides much easier (see Atlas, plate 8). In the Leiden–Orange G Papanicolaou modification this is a great advantage.

10.6.3 Physical Means to Remove Erythrocytes from Fluids

There are also physical means to remove erythrocytes from fluids.

(1) Lymphoprep gradient technique (To *et al.*, 1983).
(2) Capillary/buffy coat technique (To *et al.*, 1983; see figure 10.12).
(3) Capillary/microhaematocrit centrifuge technique (Yam and Janickla, 1983).
(4) Percoll density centrifugation (Nagasawa and Nagasawa, 1983).

10.7 Preparation of Cytological Slides from Histological Material

10.7.1 Imprint Techniques

Imprint techniques have become very popular in conjunction with, or replacing, frozen-section histology (Lee, 1982) or in the study of cell populations in biopsies of, for instance, the skin (Dracopoulou *et al.*, 1976), lymph nodes (Fernberg *et al.*, 1980) or neurological material (Gandolfi *et al.*, 1983). The freshly cut tissue surface is pressed on a slide, and air dried or immediately wet fixed.

10.7.2 Scraping Technique

The freshly cut tissue is scraped with a knife and the material thus collected is smeared.

10.7.3 Squashing Technique

The histological material is cut into pieces of 1–2 mm, and these are squashed between two slides. The slides are air dried or immediately wet fixed (see figure 10.13).

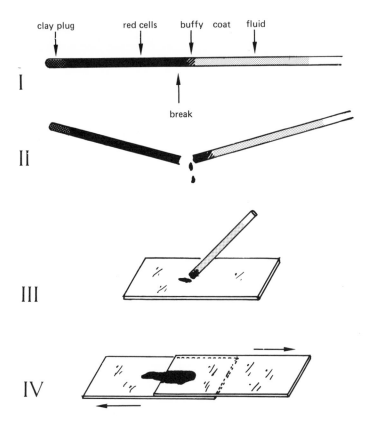

Figure 10.13 Removing erythrocytes from cell sample with the buffy coat technique. The capillary is broken at the level of the buffy coat, and smears are made of the fluid containing the cell sample. Careful *not* to smear the buffy coat itself, because it spoils cytomorphology.

10.7.4 Suspension Technique

Small biopsies ($<$ 3 mm) can be put in a suspension fluid. The container is shaken, the biopsies are removed using a tea strainer, and the cells in the suspension fluid are concentrated.

10.8 Adhesives

Pre-fixed cells have a particular tendency to adhere poorly to the slide due to the fact that the fixative has made the cell stiffer (see chapter 4). Therefore special measures can be taken to promote adherence by coating the slides. The following methods are used:

(1) coating with poly Lysine L;
(2) coating with egg albumen;
(3) coating with Apathy's syrup;
(4) coating with gelatin–chrome alum.

Coating with egg albumen has little effect, as has been shown by Nielsen *et al.* (1983) who compared different adhesives. Of those studied, gelatin–chrome alum was superior (see table 10.4). The gelatin–chrome alum coated slides are usable for weeks, and can be stored in boxes. After staining with Haematoxylin, the gelatin–chrome alum coated slides exhibit a reddish background, but the cell structures are not blurred.

10.9 Preservatives for Fluids or to Prepare Suspensions

Preservatives can be added to body fluids, urine or sputum if the material is not processed immediately. Cells from aspirates, brushes or cell pellets can also be suspended in preservatives (table 10.2). Most preservatives pre-fix the cells. All cells harvested must be post-fixed for the Papanicolaou method with 95% ethyl alcohol or with spray fixative. Some can be used as irrigation fluid. Mucolexx, containing polyethylene glycol–methanol, can also be used as preservative.

Some preservatives do *not* contain a fixative (table 10.2). The cells are then kept in a good state by putting them in a physiological environment. Culture media and solutions containing bovine albumin are used for this purpose. The cells cannot be kept for long periods of time in these preservatives, and the suspension fluids should be kept in the deep-freezer. Certain preservatives contain bacteriostatic agents (table 10.2). These are especially useful if the preservative does not contain a fixative, and when bacterial growth is a problem, as in urine.

10.9.1 Preservatives for Fluids

Preservatives added to fluid should not contain high percentages of ethyl alcohol because the cells

Table 10.4   Effectiveness of various adhesives in pre-fixed urinary cells (Nielsen *et al.*, 1983)

Coating	*Number of cells on slide (%)**	
	Median	*Range*
No adhesive	29	16–44
Egg albumen–glycerine	28	16–39
Apathy's syrup	40	31–55
Gelatin–chrome alum	96	94–99

* Number of cells as a percentage of cells on slide plus filter, 10 experiments. Similar urine specimens from the same batch of pooled urine were used in all 10 experiments.

would be stiffened too much, preventing cell adherence. Whether a preservative should be added to a fluid depends on the time between sampling and cytopreparation, and whether the high osmolality, low protein content or bacterial growth in it will damage the cells (table 10.2). Fluids with low protein contents are harmful: CSF is the best example of this (see section 10.2.6). In urine, bacterial growth is the main problem, so the urine should be pre-fixed or a preservative containing a bacterial agent should be added. In general, it is not necessary to add a preservative to coelomic fluids (see section 10.2.7). For the Papanicolaou method, it is advisable that the slides be post-fixed.

10.9.2 Preservatives for Preparing Cell Suspensions

The great advantage of suspending cell material as aspirates, sputum and brushes lies in the fact that the material is evenly distributed over the slides, and that several techniques can be performed on the cells (Boon *et al.*, 1980). In table 10.2, preservatives to be used for preparing cell suspensions are listed. Sputum must in addition be blended (10.2.3.2). For the Papanicolaou method care should be taken to post-fix the slides.

10.10 Fixatives

From the myriad fixatives used in histology, only a few are used in routine cytology (table 10.5). Some fixation methods require post-fixation. The most popular fixative is ethyl alcohol: when percentages lower than 70% are used, post-fixation is needed. Most commercial spray fixatives, as well as Saccamano's fluid and Leiden spray fixative, contain polyethylene glycol in addition to ethyl alcohol. When used as a spray, it should be removed prior to staining. The next most popular cytological fixative is methanol, which is followed in popularity by formalin.

Certain histochemical staining techniques require special fixatives, as shown in table 10.5. Smears are wet fixed either by submerging in fixation fluid or by covering with a spray fixative.

10.10.1 Wet Fixation

Wet fixation is fixation of the fresh cells with fluids, as listed in table 10.5. Especially when the smeared cells are *not* in mucoid material, care should be taken that they do not unintentionally dry. The cap of the alcohol container or spray-can should therefore be opened *prior* to taking the sample. If the material is exceedingly dry, as in vulvar smears, the slide can be pre-moistened with spray fixative. When using a spray fixative, the slide should not be closer than 15 cm to the can. The *amount* of applied spray fixative is also important: not too much so that the redundant spray fixative with cells suspended in it drips from the slide, and not too little so that the cells are not impregnated with it. It is impossible to fix cells too long, but it is certainly possible to fix cells for too short a time (shorter than 15 min). The effects of too short a fixation time are described in section 4.7. The chromatin is not crisp but 'blurred', the staining is too pale, and the nucleoli are invariably blue. If one is not certain that the slide is not properly fixed and is partly air dried, it is advisable to post-fix overnight in van der Griendt's fluid.

10.10.2 Fixation by Air Drying

Cytological slides can be air dried. The factors that influence this process are discussed in chapters 2 and 9. In short, for optimal results the cells should be in a monolayer as much as possible, and dry as fast as possible in a physiological environment, with physiological osmolarity and protein content. In aspirates this is the case, but not in coelomic fluids, CSF or urine. In these it is advisable to adjust the fluid in which the cells finally dry, or to pre-fix the cells with methanol. The air-drying techniques require much more technical skill and knowledge than the wet-fix methods, in which slides are dropped in a container with alcohol. In the latter, only too slow or too little fixation spoils the results.

Air-dried smears *must* be post-fixed with one of the fixatives listed in table 10.5. The one most commonly used is 100% methanol.

Table 10.5   Fixatives used in routine cytology

Fixatives	Post-fixation recommended	Stain
One part 37% formaldehyde, nine parts 100% ethyl alcohol	—	Papanicolaou
Neutral formalin, 10%	—	Oil Red O, Papanicolaou
Cold concentrated formalin vapour	—	Oil Red O
Cold Carnoy's	—	Methyl Green–Pyronin
Methanol, 100%*	—	Giemsa, Papanicolaou
Ethyl alcohol, 95%	—	Papanicolaou
50% ethyl alcohol: polyethylene glycol (Saccomano's fluid)	Yes, with 95% ethyl alcohol	Papanicolaou
50% methanol, 1% thymol	Yes, with 95% ethyl alcohol	Papanicolaou
Leiden spray fixative	—	Papanicolaou
Commercial spray fixatives	—	Papanicolaou
Methanol, 80%*	—	PAS
One part glycerine, five parts 50% ethyl alcohol	Yes, with 95% ethyl alcohol	Papanicolaou
13 g picric acid dissolved in 70% ethyl alcohol	—	Haematoxylin–Eosin, cell blocks
van der Griendt's fluid	Yes, with 100% ethyl alcohol 1 h or with spray fixatives	Papanicolaou

* Air-dried cells.

10.11 Various Combinations of Fixation and Staining Techniques

Although historically the Papanicolaou technique is linked to wet fixation with alcohol, the Romanowsky–Giemsa method to air drying, and the Haematoxylin–Eosin method to formalin fixation and histology, this has changed in recent times. All kinds of combinations are possible, if the cell preparation, fixation and staining are performed well. The only thing *not* to be advised is to stain air-dried smears directly with the Papanicolaou method. However, if the dry smears are incubated overnight using van der Griendt's method (1984), excellent results can be achieved with the Papanicolaou method (see Atlas, plate 8).

From figures 10.14 and 10.15, it becomes clear that the combination that should be chosen is that which gives best visualisation of a given diagnostic problem. In addition, for certain histochemical stains, special fixatives are needed (see table 10.5). Further studies for EM and SEM require special handling of the material (see next section).

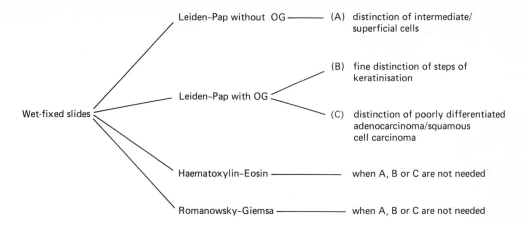

Figure 10.14 Combinations of wet-fix methods. (Note: these include fixation with ethyl alcohol, formalin and spray fixatives.)

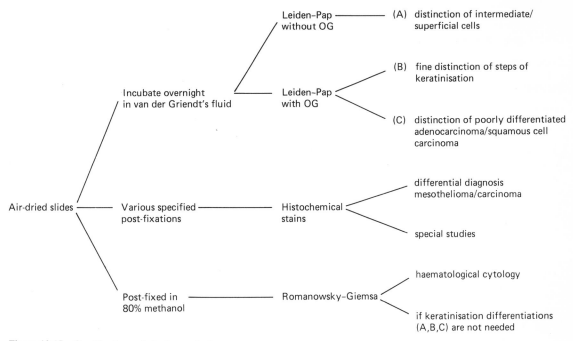

Figure 10.15 Combinations of air-dry methods.

10.12 Preparing Cytological Samples for Electron Microscopy

10.12.1 Preparation of Fine Needle Aspirates

The method according to Lindholm *et al*. (1979) is as follows:

(1) Eject cell sample directly from the needle into a small tube containing 2% glutaraldehyde in 0.1 M cacodylate buffer with 0.1 M sucrose, pH 7.4, at 4°C. The cells adhere to the fixative in sheets, which renders centrifugation unnecessary.

(2) Post-fix for 60 min in 2% osmium tetroxide buffered with 0.1 M symcollidine, pH 7.4

(3) Dehydrate in graded alcohols.

(4) Embed in Epon.

The procedure given in Mennemeyer *et al.* (1979) is as follows:

(1) Eject the cell sample directly in Mucolexx.
(2) Centrifuge at 1375 r.p.m. for 10 min, remove supernatant.
(3) Layer the sediment on Ficoll and Hypaque.
(4) Centrifuge this mixture further at 1375 r.p.m. for 10 min.
(5) Remove the cellular material and place in a capsule filled with Karnovsky's solution.
(6) Post-fix in osmium tetroxide, dehydrate in graded alcohols and embed in Epon.

10.12.2 Preparation of Alcohol-fixed Papanicolaou-stained Cells for SEM

The method according to Becker *et al.* (1981) is as follows:

(1) Select, light microscopically, the areas and cells of interest.
(2) Take black-and-white photomicrographs.
(3) Remove coverslip after soaking in xylene.
(4) Cut the slides into pieces containing the desired areas to fit them in the electron microscopy specimen chamber.
(5) Dehydrate in amyl acetate or 100% ethyl alcohol.

(6) Dry specimen by the critical point method.

10.12.3 Preparation of Glutaraldehyde-fixed, Papanicolaou-stained Slides for EM

Ruiter *et al.* (1979) use the following procedure:

(1) Smear the cell on a plastic sheet with the dimensions of commonly used object slides.
(2) Fix immediately in sodium cacodylate-buffered 1.5% glutaraldehyde for 20 min at 0–4°C.
(3) Rinse the sheet with the same buffer.
(4) Stain with the Papanicolaou method, omitting xylene.
(5) Mount in buffer.
(6) Select cells and areas of interest and mark them.
(7) Post-fix for 15 min at 0–5°C in sodium cacodylate-buffered 2% osmium tetroxide.
(8) Dehydrate in graded alcohols and infiltrate with a propylene oxide epoxy resin mixture; avoid drying of the cells during the procedure. Let the mixture evaporate overnight.
(9) Wipe the epoxy resin from the plastic sheet except for the marked areas.
(10) Place a gelatin embedding capsule filled with Epon 81^2 face down on the marked area and incubate for 24 h at 70°C.
(11) Remove the plastic sheet after polymerisation: the marked cells are then incorporated in the Epon block and thus lifted from the sheet.

Appendix 1

Fixatives*

Immersion Fixatives

The cytological slide is immersed in a solution containing fixative. For optimal results, the slide should remain immersed for at least 15 min: longer fixation times do not change the cell images.

95% Alcohol

Slides are to be immersed in the alcohol (15–30 min is sufficient)

Equivalents for 95% Alcohol

100% Methanol
80% Propanol
80% Isopropanol (can also be used as a spray fixative)

Ethanol–Ether

Mix 95% alcohol and ether in equal quantities

Carnoy's Fixative

95% Alcohol	60 ml
Chloroform	30 ml
Glacial acetic acid	10 ml

Methacarn

Methanol	60 ml
Chloroform	30 ml
Glacial acetic acid	1 ml

Propanol–Glycerine

Isopropanol	80 ml
Glycerine	40 ml

10% Neutral Buffered Formalin

37–40% Formaldehyde solution	100 ml
Water	900 ml
Acid sodium phosphate, monohydrate	4.0 g
Anhydrous disodium phosphate	6.5 g

4% Saline Formalin

37–40% Formaldehyde solution	5 ml
NaCl	0.9 g
Water	90 ml

The mixture haemolyses erythrocytes. It can be used for pre-fixation but also for post-fixation of air-dried smears before applying the Papanicolaou staining method (van der Griendt, 1984, pers. comm.).

* In all cases where 'alcohol' or 'ethanol' is mentioned, ethyl alcohol is meant.

Neutral Formalin

37–40% Formaldehyde solution	10 ml
Water	90 ml
Calcium carbonate	about 1 g

Add calcium carbonate in excess.

Bouin's Fluid

1.2% (= saturated) picric acid	75 ml
36–40% Formalin	25 ml
Glacial acetic acid	5 ml

The amount of formalin may be reduced to 10 ml.

Formalin Vapour Fixation

Place 1–2 ml of formalin solution of required strength in Coplin jar. Immediately after taking sample, put slide in jar with cell-coated side uppermost and tightly cover the jar. Leave the slide in the jar for required length of time depending on the procedure which is to follow.

20–25% Glutaraldehyde

For prolonged storage the solutions should be kept at a low temperature (4°C or lower) and low pH.

2% Formaldehyde and 2% glutaraldehyde

2% Formaldehyde	25 ml
2% Glutaraldehyde	25 ml
NaCl	4.5 g

Osmium tetroxide

Osmium tetroxide	1–2 g
Water	100 ml

Slides are to be kept in the solution overnight.

Zenker's Fixative

$K_2Cr_2O_7$	2.5 g
$HgCl_2$	5–8 g
Distilled water	100 ml

Glacial acetic acid, 5% by volume to be added at the time of use. Dissolve salts in water, heating gently.

Spray Fixatives

Hair-spray

Any hair-spray with a high alcohol content and some lanolin or oil will do.

Polyethylene Glycol

95% Alcohol	500 ml
Ether	500 ml
Polyethylene glycol (Carbowax)	50 g

NB Ether may be omitted and instead 100 ml 95% alcohol may be used.

Polyethylene glycol is available in various grades of viscosity, indicated by a number. We prefer number 400, which has the consistency of a syrupy liquid. Polyethylene glycol 1500 is like vaseline and stays on the slide without drying out. For that reason it may be preferred but it also means that the cell film plus coating can be damaged more easily.

Leiden Spray Fixative

96% Alcohol	700 ml
Acetone	200 ml
Polyethylene glycol (see note above)	70 ml

NB Perfume bottles can be used for spraying the fixatives on the slides. It is very important to spray from the correct distance; 25–30 cm gives optimal results.

For sputum, 70% alcohol is recommended. The acetone can also be replaced by alcohol. These fixatives can also be used as immersion fixatives. If the material is bloody, it is advisable to add 50 ml glacial acetic acid to 1000 ml fixative.

Air Drying

Some methods (like the Romanowsky–Giemsa staining method or the colouring of lipids) have best results after air drying. This is usually followed by some form of post-fixation.

Fixatives for Cell Suspensions

Saccomano's Fixative

50% Alcohol	100 ml
Carbowax 1540	2 g

Esposti's Fluid

Methanol	225 ml
Purified water	225 ml
Glacial acetic acid	50 ml

Alcohol Solutions

50% Alcohol for cell suspensions from all body sites except gastric
70% Alcohol for sputum, bronchial aspirates and washings
90% Alcohol for gastric, bronchial and other saline washings

Formalin Vapour Fixation

See preceding section on immersion fixatives. The fixation time can be shortened to 1 min when the procedure is done in the microwave oven at high energy level.

Preserving Fluids

Cell suspensions can be kept for a limited period of time without fixation. To preserve the cells, the medium should have physiological salt concentration and some essential nutrients, preferably with an anti-mould agent added. There are a number of commercial preserving fluids on the market, of which the Minimum Essential Medium (MEM) (Eagle) is perhaps best known. It contains glutamine, salts, and sodium azide as preservative.

Delft Suspension Medium

MEM	500 ml
6% Foetal calf serum	30 ml
1% Glutamine	5 ml
200 U/ml Penicillin	1 ml
0.2 mg/ml Streptomycin	1 ml
0.5 μg/ml Fungizone	1 ml

pH should be 7.18.

Phosphate-buffered Saline

NaCl	87.5 g
$Na_2HPO_4.12H_2O$	23.25 g
KH_2PO_4	2.15 g

Dissolve in 1 litre distilled water. pH is 7.4.

Preservative According to Marsan *et al.* (1982)

0.9% NaCl	249 ml
22% bovine albumin	1 ml

Adjust to pH 7.5.

Formol–saline (van der Griendt's Fluid)

Physiological saline	1000 ml
40% Neutral formalin	2.5 ml

This solution haemolyses erythrocytes. It can also

be used for incubating air-dried smears prior to staining with the Papanicolaou method (see section 10.4.2).

Clearing Agents

As the name implies, clearing agents should clear the cell preparation. This is done after dehydration with increasing concentrations of alcohol.

Xylene is a well established clearing agent. It is a very unpleasant chemical to have around in the laboratory. It does not mix with water. If there should be some water left on the cell preparation (for instance if the last dehydration bath is not 100% alcohol), this will show as a white film on slide.

Tertiary butanol is less demanding. It is less unpleasant to use and does not show a white film if water is left in the preparation. Its disadvantage is that its melting point is higher than the usual laboratory temperature, so that it needs either heating or to be mixed with a solvent to make it fluid. Since the former possibility is not always wanted, the latter is the better choice. After melting down the solid tertiary butanol it can be mixed with 96% alcohol (15:1) after which it remains liquid (Drijver and Boon, 1983a).

Mounting Media

Mounting media have a double task. They should:

(1) smooth out the cell preparation, i.e. make the surface of the preparation even so that there are no gaps left between cells and cover glass;
(2) glue together cell preparation and cover glass so that the slide can be kept without the preparation being damaged and air getting in.

Mounting media should have a number of properties in order to make them suitable for the purpose for which they are intended. First, they should have a refractive index similar to that of glass so that the image does not become distorted. The RI should not be the same as that of the tissue components. If it were the same, unstained or slightly stained cells would be difficult to see.

Secondly, the mounting media should not influence the colour of the cells by either fading or 'bleeding' (this is the running of the dye from its original site). For this reason it is important to know the acidity of the medium.

Thirdly, the dye or dyes should not dissolve in the mounting medium. For that reason, only water-based media can be used in fat-stained preparations (see below).

The requirements mentioned above are mostly met by the resins. These can be either natural, semisynthetic or synthetic. The natural resins (Canada balsam and cedar oil) are slightly acid and tend to fade basic dyes, especially cedar oil. The semisynthetic resins (like Euparal and Diaphane) are also slightly acid. Euparal can fade Haematoxylin but basic aniline and the Romanowsky–Giemsa stains are well preserved. There are a number of synthetic resins on the market. They are all quite neutral and have a wide range of applications.

Cells which are stained for fat cannot be mounted in any of the above-mentioned media since the dye would dissolve in them. For this kind of cell preparation, water-based media should be used, with gelatin or gum as hardener. Slides which need not be kept permanently can be mounted in just glycerine for as long as necessary or practicable.

Glycerine–Gelatin

Gelatin	40 g
Water	210 ml
Glycerine	120 ml

Soak the gelatin in the water for 2 h. Add the glycerine, stirring all the time. The solution should be kept at 0–5°C and melted as needed. To prevent mould, 50 mg Merthiolate, 100 mg thymol or 5 ml melted phenol may be added. Phenol may have a damaging effect on Haematoxylin.

Apathy's Gum Syrup

Gum arabic (acacia)	50 g
Cane sugar	50 g
Water	100 ml

Mix the ingredients while heating gently. A preservative may be added as in glycerine–gelatin. The suger is added to raise the refractive index of the gum, which is very low. Other sugars may be used. The important thing is to use a sugar that does not crystallise.

Blueing Solutions

(1) Scott's tap water substitute (commercially available).
(2) Lithium carbonate solution (three drops of a saturated lithium carbonate solution in 100 ml water).
(3) Water to which a few drops of ammonia have been added.

NB Blueing solutions should have a pH of about 8.

Appendix 2

Staining Methods*

In this appendix we have collected together a number of staining methods. The Papanicolaou and Romanowsky–Giemsa methods are represented by various modifications. In addition, certain methods in which Haematoxylin is substituted by other dyes are given. Also some methods are included for specific dyeing of nuclear and cytoplasmic components and pigments, and bacteria.

Some general remarks must now be made. First, unless otherwise mentioned, 'bring to water' or 'hydrate' means that after fixation the slides should be brought through graded baths of alcohol (96, 70, 50%) to water. Secondly, 'dehydrate' means bringing the slides to absolute alcohol through graded baths of alcohol. Unfortunately, the pH of the dye bath is rarely mentioned in staining methods described elsewhere. We have therefore included details wherever possible. We recommend measuring the pH of every dye solution and keeping a check on it if the solution is retained for a longer period or used for the staining of a large number of slides.

The recipes of the solutions needed for a staining procedure are either given under the heading 'Solutions needed' or marked with a superscript number, in which case they can be found in appendix 3 under the appropriate staining solution.

Not all staining procedures need a specific fixation method. In that case we have recommended spray fixative or 96% alcohol since these are the methods most commonly used. This does not necessarily mean that other routinely used fixatives cannot be used or tried.

---

* Where 'alcohol' is mentioned, ethyl alcohol (ethanol) is meant.

1

Papanicolaou's Staining Method including Orange G (Leiden Modification)

Purpose

General nuclear and cytoplasmic stain.

Solutions Needed

Cole's Haematoxylin[5] (pH 6.5)
EA – Leiden[14] (pH 4.6)
Orange G – Leiden[19]
Tertiary butanol/95% alcohol, 15 : 1

Fixation

Leiden spray fixative.

Procedure

(1) Place slides in 50% alcohol to remove polyethylene glycol for at least an hour but longer if more convenient (overnight, for instance).
(2) Rinse in tap water, 15 dips.
(3) Stain in Cole's Haematoxylin for 2 min.
(4) Rinse in tap water till water runs clear.
(5) Dip in 50% alcohol, 15 dips.
(6) Stain in Orange G – Leiden, 1 min.
(7) Dip in two changes of 50% alcohol, 15 dips each.
(8) Stain in EA – Leiden, 2 min.
(9) Dip in 50% alcohol, 15 dips.
(10) Dehydrate.
(11) Clear in tertiary butanol/alcohol.
(12) Mount.

Result

Nuclei blue to bluish purple.
Nucleoli blue or red.
Cytoplasm in superficial squamous cells pink, in intermediate cells turquoise blue.
Nuclei in lymphocytes and leucocytes blue.
Erythrocytes pink – orange.
Highly keratinised cells orange.

2

Papanicolaou's Staining Method without Orange G (Leiden Modification)

Purpose

General nuclear and cytoplasmic stain.

Solutions Needed

Cole's Haematoxylin[5] (pH 6.5)
EA – Leiden[14] (pH 4.6)
Tertiary butanol/95% alcohol, 15 : 1

Fixation

Leiden spray fixative.

Procedure

(1) Place slides in 50% alcohol to remove polyethylene glycol for at least an hour but longer if more convenient (overnight, for instance).
(2) Rinse in tap water, 15 dips.
(3) Stain in Cole's Haematoxylin for 2 min.
(4) Rinse in tap water till water runs clear.
(5) Dip in 50% alcohol, 15 dips.
(6) Stain in EA – Leiden for 2 min.
(7) Dip in 50% alcohol, 15 dips.
(8) Dehydrate.
(9) Clear in tertiary butanol/alcohol for at most 5 min.
(10) Air-dry if slides are to be coverslipped automatically; if covered by hand, do so straight from the clearing bath without drying.

Result

Nuclei blue to bluish purple.
Nucleoli blue or red.
Cytoplasm in superficial squamous cells pink, in intermediate cells turquoise blue.
Nuclei in lymphocytes and leucocytes blue.
Erythrocytes pink – red.

3

Papanicolaou's Staining Method (following Koss, 1979)

Purpose

General nuclear and cytoplasmic stain.

Solutions Needed

Harris's Haematoxylin[2]
0.1% Hydrochloric acid
Saturated solution of lithium carbonate
Orange G[18]
EA−35 or EA−65[15]

Fixation

Spray fixative or 96% alcohol.

Procedure

(1) Leave in 95% alcohol to remove polyethylene glycol for at least an hour but longer if preferrred.
(2) Hydrate.
(3) Distilled water, 15 dips.
(4) Harris's Haematoxylin, 6 min.
(5) Rinse in running tap water till water runs clear.
(6) Dip in distilled water, 15 dips.
(7) Stand in distilled water, 10 min.
(8) Differentiate in 0.1% HCl till colour of slide turns red (check!).
(9) Dip in distilled water, 15 dips.
(10) Lithium carbonate, 2 min.
(11) Dehydrate.
(12) Stain in Orange G, 2 min.
(13) Dip in 96% alcohol, 15 dips.
(14) Stain in EA−35 or 65, 4 min.
(15) Dip in two changes of 96% alcohol, 15 dips each.
(16) Leave in 100% alcohol, 2 min.
(17) Clear in xylene.
(18) Mount.

Result

Nuclei blue.
Nucleoli dark blue or red.
Cytoplasm of superficial squamous cells pink, of intermediate cells turquoise-blue or light blue, depending on EA used.
Erythrocytes red.
Nuclei of leucocytes and lymphocytes dark blue.

4

Papanicolaou's Staining Method (Gill's Modification)

Purpose

General nuclear and cytoplasmic stain.

Solutions Needed

Gill's Haematoxylin[6]
Gill's Orange G−6[17]
Scott's tap water substitute if pH of local tap water is below 7.0
Gill's EA[16]

Fixation

95% Alcohol, at least 15 min.

Procedure

(1) Dip in two changes of water, 10 dips each.
(2) Stain in Gill's Haematoxylin, 2 min.
(3) Dip in two changes of water, 10 dips each.
(4) Leave in Scott's tap water substitute, 1 min.
(5) Dip in two changes of water, 10 dips each.
(6) Dip in two changes of 95% alcohol, 10 dips each.
(7) Stain in Gill's OG−6, 1 min.
(8) Dip in three changes of 95% alcohol, 10 dips each.
(9) Stain in Gill's EA, 4−10 min.
(10) Dip in three changes of 95% alcohol, 10 dips each.
(11) Dip in 100% alcohol, 10 dips.
(12) Clear in xylene and mount.

Result

Nuclei blue.
Cytoplasm of superficial squamous cells pink, of intermediate squamous cells turquoise blue.
Nucleoli dark blue.
Nuclei of leucocytes and lymphocytes dark blue.
Keratinised cells orange.
Erythrocytes pink−red.

5

Rakoff's Staining Method (Rakoff, 1960)

Purpose

In-office hormonal evaluation of gynaecological samples.

Solutions Needed

Rakoff's stain[21]
Saline solution (0.9% NaCl)

Procedure

(1) Make swab with cotton tip dipped in saline.
(2) Drop swab in test tube containing 1–2 ml saline.
(3) Place 3 drops of stain into test tube and stir solution with swab.
(4) Transfer 1 or 2 drops to slide and coverslip.

Result

Cytoplasm of superficial cells red, of intermediate cells green.
Vesicular nuclei show sharply stained margins but are otherwise pale.

6

Shorr's Staining Method

Purpose

For hormonal assessment.

Solutions Needed

Shorr's stain[20]

Fixation

95% Alcohol.

Procedure

(1) Stain in Shorr's solution, 1 min.
(2) Dehydrate.
(3) Clear in xylene.

Result

Cytoplasm of superficial squamous cells orange–pink, of intermediate cells green.

7

Szczepanik's Quick Staining Method (Szczepanik, 1978) for Papanicolaou Method

Purpose

To stain cell sample during consultation.

Solutions Needed

Harris's Haematoxylin[2]
Isopropyl alcohol
EA-31 (Merck No. 9271 or 9272)
Water at 40–70°C

Fixation

Spray fixative.

Procedure

(1) Dip in two changes of propanol, 1 dip each.
(2) Stain in Haematoxylin, 70 s.
(3) Rinse in water (40–70°C), 5 s.
(4) Dip in propanol, 1 s.
(5) Stain in EA, *c.* 38 s.
(6) Clear in xylene, two changes, 1 s each.

Result

Nuclei blue.
Nucleoli dark blue.
Cytoplasm of superficial squamous cells pink, of intermediate cells light blue.
Erythrocytes pink.
Nuclei of lymphocytes and leucocytes dark blue.

8

Delafield's Haematoxylin–Eosin Y Method

Purpose

General nuclear and cytoplasmic stain.

Solutions Needed

Delafield's Haematoxylin[1]
Acid alcohol: 3 drops of concentrated HCl in 50 ml 95% alcohol 0.1–0.5% Eosin Y in 25% alcohol

Fixation

Spray fixative or 96% alcohol.

Procedure

(1) Dissolve polyethylene glycol in 50% alcohol.
(2) Dip in water, 15 dips.
(3) Stain in Haematoxylin, 15 min.
(4) Rinse in tap water till water runs clear.
(5) Stand in tap water for 10 min.
(6) If slides are still very blue, add 2 drops of acid alcohol.
(7) Stain in Eosin Y solution.
(8) Rinse in distilled water.
(9) Dehydrate.
(10) Clear in xylene.
(11) Mount.

Result

Nuclei dark blue.
Cytoplasm pink–red.

9

Gallocyanin Chrome Alum (Sandritter *et al.*, 1966)

Purpose

To determine the total nucleic acid content of the nucleus.

Solutions Needed

Gallocyanin chrome alum,[10] pH 1.64

Fixation

96% Alcohol.

Procedure

(1) Stain in Gallocyanin solution for 48 h.
(2) Differentiate in dehydrating alcohols.
(3) Clear in xylene and mount.

Result

Nuclei blue–black.
RNA-rich sites in cytoplasm blue–black.

10

Coelestine Blue (Gray *et al.*, 1956; Yasumatsu, 1977)

Purpose

To be used as a general nuclear stain as a Haematoxylin substitute.

Solutions Needed

Coelestine Blue solution according to Gray *et al.*[9] or Yasumatsu[8]

Fixation

Any fixative can be used which is also suitable for Haematoxylin staining.

Procedure

(1) Hydrate slides.
(2) Stain in Coelestine Blue solution, 1 min to 1 h.
(3) Rinse in water.
(4) Dehydrate, clear and mount.

Result

Nuclei blue (the colour is an intermediate between the blue from Haematoxylin and that from Methylene Blue).

11

Nuclear Staining with Carminic Acid (Mayer, 1891)

Purpose

To be used as pure nuclear dye.

Solutions Needed

Carminic Acid solution[7]
0.1–1% Potassium alum solution in water

Fixation

Spray fixative or any other routine fixative.

Procedure

(1) Bring slides to water.
(2) Stain in Carminic Acid solution, 15 min.
(3) Rinse in water, 2–3 min.
(4) If needed, differentiate briefly in potassium alum.
(5) Rinse thoroughly in water.
(6) Dehydrate.
(7) Mount.

Result

Nuclei red, very clearly defined.

12

The Feulgen Reaction

Purpose

To show DNA.

Solutions Needed

Schiff's reagent[13]
1 N HCl at 60°C
0.05 M Metabisulphite
0.01% Fast Green FCF in 95% alcohol if wanted

Fixation

Any fixative may be used.

Procedure

(1) Hydrolyse in HCl for 10 min.
(2) Leave in Schiff's reagent for 10 min.
(3) Wash in three successive baths of metabisulphite for 2 min each.
(4) Wash in running water.
(5) Counterstain, if wanted, for a few seconds.
(6) Dehydrate, clear and mount.

Result

DNA-rich sites red.
Basic cell components green.

13

Combined Feulgen–Naphthol Yellow S Method (Gaub *et al.*, 1975)

Purpose

To demonstrate DNA and basic proteins for quantitative measurement.

Solutions Needed

Schiff's reagent[13]
5 N HCl at 22°C
0.05 M Metabisulphite
0.1% Naphthol Yellow S in 1% acetic acid, pH 2.8
1% Acetic acid
Tertiary butanol/absolute alcohol 15 : 1

Fixation

10% Neutral formalin is preferred, but it is also possible to use alcohol–acetone.

Procedure

(1) Hydrolyse in HCl for 60 min at 22°C.
(2) Rinse in metabisulphite.
(3) Rinse in distilled water.
(4) Stain in Naphthol Yellow S for 30 min.
(5) Destain in 1% acetic acid, 3 baths, 0.5 min each.
(6) Clear in tertiary butanol–alcohol, 3 baths.
(7) Clear in xylene 5 min.
(8) Mount.

Result

DNA-rich sites blue.
Basic proteins yellow.

14

Methyl Green–Pyronin Y (Taft, 1951; Clark, 1973)

Purpose

To stain DNA and RNA differentially.

Solutions Needed

Methyl Green–Pyronin Y solution[11]
Tertiary butanol–alcohol, 3 : 1

Fixation

Carnoy.

Procedure

(1) Bring slides to water.
(2) Stain in Methyl Green–Pyronin Y solution, 3–5 min.
(3) Rinse in distilled water.
(4) Blot dry with smooth filter paper.
(5) Before completely dry, differentiate for at least 2 min in tertiary butanol–alcohol mixture.
(6) Clear in two changes of xylene, 5 min each.
(7) Mount in synthetic resin.

Result

Chromatin blue–green.
Nucleoli rose.
Cytoplasm granules dark rose.
Cytoplasm of plasma cells dark rose, occasionally almost purple.
Mast cell granules refractile, orange–red.

15

Methyl Green–Pyronin Y (Kurnick, 1952)

Purpose

To stain DNA and RNA differentially.

Solutions Needed

0.2% Methyl Green in water (chloroform extracted)
Saturated solution of Pyronin Y in acetone
n-Butyl alcohol
Cedar oil
Dilute solutions of Pyronin Y may be used if a more delicate stain is wanted

Fixation

Carnoy.

Procedure

(1) Bring slides to water.
(2) Stain in Methyl Green solution, 6 min.
(3) Blot dry.
(4) Differentiate in two changes of *n*-butyl alcohol, 2–3 min each.
(5) Stain in Pyronin Y solution, 30–90 s.
(6) Transfer directly to cedar oil for clearing.
(7) Further clear in xylene.
(8) Mount in Permount.

Result

Nucleoli pink.
Cytoplasm red.
Chromatin bright green.
Erythrocytes brown.

16

Pontacyl Dark Green B (Bedrick, 1970)

Purpose

To stain nucleoli differentially.

Solutions Needed

2% Pontacyl Dark Green B in water; to 100 ml, 2N HCl (2 ml) is added.

Fixation

10% Formalin.

Procedure

(1) Bring slides to water.
(2) Stain in Pontacyl Dark Green B solution, 3 min.
(3) Rinse in water.
(4) Dehydrate and mount.

Result

Nucleoli in malignant cells more intense green than those of benign cells.
Good differentiation of cellular structures.

17

Love's Toluidine Blue Molybdate Method (Love *et al.*, 1973)

Purpose

To stain nucleolini.

Solutions Needed

15 mg% Toluidine Blue in McIlvaine's buffer[12]
10% Trichloroacetic acid
Formol sublimate (6% aqueous mercuric chloride:
formaldehyde USP, 9 : 1)
Lugol's iodine[12,22]
0.2 M Aqueous sodium thiosulphate
1 mg% Deoxyribonuclease in Tris-hydroxymethyl
aminomethane buffer[12]
15% Aqueous ammonium molybdate[12]
Tertiary butanol

Fixation

Do not allow smears to dry! 10% Trichloroacetic
acid, 10 min at room temperature followed by a
rinse in a dish of tap water, 5 s. Then fix in formol
sublimate, 5 min exactly.

Procedure

(1) Rinse slides in running tap water.
(2) Remove sublimate in Lugol's iodine, 5 min.
(3) Wash off Lugol's with tap water.
(4) Immerse in sodium thiosulphate, 5 min.
(5) Rinse in running tap water, 5 min.
(6) Leave in DNA-ase solution, 60 min at 37°C.
(7) Stain in Toluidine Blue, 120 min.
(8) Rinse in running tap water, 10 s exactly.
(9) Immerse in ammonium molybdate, 7 min
    exactly.
(10) Rinse in running tap water, 10 s exactly.
(11) Drain off excess water.
(12) Dehydrate in tertiary butanol at a tem-
     perature just above 25.5°C (= melting point).
(13) Clear in xylene and mount in Permount.

Result

Nucleolini purple.
Body of the nucleolus pale blue.

18

Amido Black 10B (Synonym Pontacyl Blue Black SX) (Mundkur and Greenwood, 1968)

Purpose

To stain nucleoli differentially in lymph nodes in
Hodgkin's disease but can also be used for other
material.

Solutions Needed

2% Phosphomolybdic acid in water
1% Amido Black 10B in water

Fixation

Carnoy's ethanol, chloroform, acetic acid, 6 : 3 : 1,
4 h.

Procedure

(1) Bring slides to water.
(2) Mordant in phosphomolybdic acid, 10 min.
(3) Rinse in running tap water, 3 min.
(4) Stain in Amido Black 10B solution, 15 min.
(5) Rinse in running tap water, 3 min.
(6) Rinse in 30% alcohol, 15 min.
(7) Rinse in 50% alcohol, 15 min.
(8) Rinse in 70% alcohol, 5 min.
(9) Rinse in 80% alcohol, 5 min.
(10) Leave overnight in 95% alcohol (shorter times
     are possible but minimum time is not
     specified).
(11) Final dehydration in 100% alcohol, 15 min.
(12) Clear in toluene and mount (Permount).

Result

Nucleoli blue.
Cytoplasm very light blue.

NB A cytoplasmic counterstain may be used if
wanted.

19

Cuprolinic Blue in the Presence of Magnesium Chloride (Mendelson *et al.*, 1983)

Purpose

To stain nuclear and cytoplasmic RNA.

Solutions Needed

0.1% Cuprolinic Blue in 25 mM acetate buffer with 1 M $MgCl_2$, pH 5.6
25 mM acetate buffer with 1 M $MgCl_2$, pH 5.6
Glycerine–gelatin, if wanted (see appendix 1).

Fixation

Air drying and post-fixing in modified Carnoy (ethanol–chloroform–acetic acid, 60 : 30 : 5).

Procedure

(1) Bring slides to water.
(2) Stain in Cuprolinic Blue solution, 60 min, at room temperature.
(3) Rinse in several changes of the dye solvent (i.e. acetate buffer with $MgCl_2$), 15 min each.
(4) Rinse in 3 changes of distilled water, 2 min each.
(5a) Dehydrate and mount in Euparol, *or* (5b) air-dry and mount in glycerine jelly.

NB The same procedure can be used if both DNA and RNA are to be stained. In that case the acetate buffer should not contain $MgCl_2$.

Result

RNA in cytoplasm blue.
Nucleoli blue.

20

Selective Staining of RNA with a Basic Dye and a Cationic Surfactant (Bennion *et al.*, 1975)

Purpose

To show RNA-rich sites in cytoplasm and nuclei.

Solutions Needed

0.1% Azure A or Toluidine Blue and 1% Hyamine 2389 in M/10 phosphate buffer, pH 7.0 (Hyamine 2389 is available commercially. It contains approximately 50% methyldodecylbenzyltrimethyl ammonium chloride in aqueous solution.)
Synthetic resin

Fixation

10% Buffered neutral formalin preferred; absolute alcohol and acetone also possible.

Procedure

(1) Bring slides to water.
(2) Stain in Azure A or Toluidine Blue solutions, 30 min.
(3) Rinse in water, blot or air dry.
(4) Mount in synthetic resin (Polymount, Searl Scientific).

Result

Basophilic cytoplasm, like that of exocrine pancreas, plasma cells, Nissl bodies and nucleoli, stain purple.
Chromatin stains pale blue.

21

Guard's Method for Sex Chromatin (Guard, 1959 detailed in Koss, 1979)

Purpose

To demonstrate Barr bodies in buccal smears.

Solutions Needed

Biebrich Scarlet solution (appendix 4)
Fast Green solution (appendix 4)
Harris's Haematoxylin[2]

Fixation

Immediately in 95% alcohol.

Procedure

(1) Transfer slide to 70% alcohol, 2 min.
(2) Stain in Biebrich Scarlet, 2 min. (See Note 1.)
(3) Rinse in 50% alcohol.
(4) Differentiate in Fast Green solution, 1 to 4 h. (See Note 2.)
(5) Leave in 50% alcohol, 5 min.
(6) Dehydrate.
(7) Clear in three changes of xylene, 2 min each.
(8) Mount.

Result

Sex chromatin (Barr body) red.
Background green.

Note 1: If Haematoxylin is to be included, this should be done before staining with Biebrich Scarlet. The slides then go immediately from the Haematoxylin bath into the Biebrich Scarlet solution. The staining results are then as follows:
Sex chromatin red.
Nuclear chromatin blue.
Cytoplasm green.

Note 2: Differentiation needs to be checked microscopically at hourly intervals. When the cytoplasm of all cells is green and nuclei are vesicular, the slide is ready. Pyknotic nuclei will not differentiate and are stained red.

22

Best's Method (Luna, 1968).

Purpose

To stain glycogen.

Solutions Needed

Best's Carmine solution[23]
Any progressive Haematoxylin solution (Cole's[5] or Mayer's[3])
Differentiating solution: 100% alcohol 20 ml, methanol 10 ml, distilled water 25 ml

Fixation

Carnoy's or formol–alcohol.

Procedure

(1) Bring slides to distilled water.
(2) Stain in Haematoxylin, 15 min.
(3) Rinse in running water, 15 min.
(4) Stain in Carmine working solution, 30 min.
(5) Leave in differentiating solution, few seconds.
(6) Rinse quickly in 80% alcohol.
(7) Dehydrate.
(8) Clear in xylene.
(9) Mount in Permount or Histoclad.

Result

Glycogen pink or red.
Nuclei blue.

23

24

Periodic Acid – Schiff Reaction

Purpose

Demonstration of glycogen.

Solutions Needed

Schiff's reagent[13]
1% Periodic acid (H_5IO_6) in water.
0.52% $NaHSO_3$ (0.05 M)
Weigert's[4] or Mayer's[3] Haematoxylin

Fixation

Any fixative may be used.

Procedure

(1) If air-dried, post-fix smear in 70% alcohol for 10 min. If spray-fixed, dissolve polyethylene in water as usual.
(2) Rinse in water.
(3) Oxidise in periodic acid for 5 min.
(4) Rinse in distilled water.
(5) Leave in Schiff's reagent for 15 min.
(6) Pass directly to 3 successive baths of $NaHSO_3$ for 2 min each.
(7) Rinse in running tap water for 10 min each.
(8) Counterstain with Haematoxylin if wanted (Weigert's or Mayer's).
(9) Rinse in running tap water for 10 min.
(10) Dehydrate.
(11) Clear in xylene and mount.

Result

Glycogen-rich sites red
Nuclei blue or black depending on Haematoxylin used.
Mucin red – purple or violet.
Cytoplasm grey or pale blue.

Periodic Acid Schiff Reaction in Combination with Amylase and Dimedone

Purpose

The selective staining of glycogen, mucoproteins and glycoproteins.

Solutions Needed

Schiff's reagent
2% Periodic acid in water
5% Dimedone† in 96% alcohol (Dimedone (5,5-dimethylcyclohexane-3,dione) is a commercial preparation for the blocking of aldehydes.)
Amylase: saliva in phosphate-buffered saline (1 : 1), pH 0.0

Fixation

Any fixative may be used.

Procedure

The staining procedure is the same as in Method 23.
   Amylase treatment: The cells are incubated for 30 min at 37°C and then stained.
   Dimedone treatment: The cells are incubated for 3 h at 60°C.

Result

PAS only: all PAS-positive material is stained.
PAS in combination with amylase: mucoproteins and glycoproteins are stained, glycogen remains unstained.
PAS in combination with dimedone: glycogen and mucin are stained.
PAS in combination with amylase and dimedone: only mucin is stained.

25

Alcian Blue Method for Mucosubstances, pH 2.5 (Luna, 1968)

Purpose

To stain polysaccharides in mucin.

Fixation

10% Buffered neutral formalin.

Solutions Needed

1% Alcian Blue 8GX in 3% acetic acid, adjusted to pH 2.5
3% Acetic acid in distilled water
0.1% Kernechtrot in 5% aluminium sulphate

Procedure

(1) Bring slides to distilled water.
(2) Mordant in 3% acetic acid, 3 min.
(3) Stain in Alcian Blue, 30 min.
(4) Rinse in running tap water for 10 min.
(5) Rinse in distilled water.
(6) Counterstain in Kernechtrot, 5 min.
(7) Rinse in running tap water, 1 min.
(8) Dehydrate and clear in xylene.
(9) Mount.

Result

At this pH, weakly acidic sulphated mucosubstances, hyaluronic acid and sialomucins stain dark blue.
Nuclei red.

26

Sudan Black B (Lillie and Fulmer, 1976)

Purpose

To demonstrate fat in cytoplasm.

Solutions Needed

Sudan Black B solution[25]
Propylene glycol
Glycerine jelly (see appendix 1)

Fixation

Formalin.

Procedure

(1) Wash in distilled water, 2–5 min.
(2) Dehydrate in propylene glycol while moving slides, 3–5 min.
(3) Stain in Sudan Black B solution, agitating occasionally, 5–7 min.
(4) Differentiate in 85% propylene glycol, 2–3 min.
(5) Rinse in distilled water, 3–5 min.
(6) Counterstain with nuclear stain if wanted.
(7) Rinse in tap water, two changes.
(8) Mount in glycerine jelly.

Result

Neutral fats greenish black.
Myelin greenish black.
Mitochondria greenish black.
Other lipids greenish black.
Cytoplasm unstained.

27

Oil Red O (ORO) (Luna, 1968)

Purpose

To demonstrate fat in cytoplasm.

Solutions Needed

100% Propylene glycol
60% Propylene glycol in distilled water
0.5% Oil Red O solution in 100% propylene glycol[24]
Haematoxylin, either Harris's or Mayer's
Dilute acid, e.g. 0.1% HCl in water
Blueing solution if pH of tap water is below 7.0
Glycerine – gelatin (see appendix 1)

Fixation

Air drying.

Procedure

(1) Post fix in 10% buffered neutral formalin (see appendix 1).
(2) Rinse in distilled water.
(3) Leave in 100% propylene glycol, 2 min.
(4) Stain in Oil Red O, 10 min.
(5) Differentiate in 60% propylene glycol, 1 min.
(6) Rinse in distilled water.
(7) Stain in Haematoxylin, 6 min.
(8) Rinse in running tap water for 10 min.
(9) Mount in glycerine jelly.

Result

Fat red.
Nuclei blue.

28

Gomori's Silver Method (Heckner Modification)

Purpose

Demonstration of fibres.

Solutions Needed

0.5% Potassium permanganate
1% Potassium metabisulphite
2% Ammonium sulphate
Gomori's silver solution[26]
4% Formalin
0.5% Gold chloride
5% Sodium thiosulphate
2% Ferrous ammonium sulphate
Giemsa stain

Fixation

Air drying and post-fixing in methanol.

Procedure

(1) Leave slides in potassium permanganate for 1–2 min.
(2) Rinse in tap water, 5 min.
(3) Destain in potassium metabisulphite for 1 min.
(4) Rinse in tap water for 5 min.
(5) Leave in ferrous ammonium sulphate, 30 s.
(6) Rinse in distilled water, 2 baths, 2 min each.
(7) Leave slide to dry.
(8) Stain in Gomori's solution in Coplin jar, 1 min.
(9) Quickly rinse in distilled water in Coplin jar, 5 s.
(10) Develop in 4% formalin in Coplin jar, 5 min.
(11) Rinse in tap water in Coplin jar, 5 min.
(12) Leave in gold chloride in Coplin jar, 5 min.
(13) Rinse in distilled water.
(14) Destain in potassium metabisulphite in Coplin jar, 1 min.
(15) Fix in sodium thiosulphate, 1 min.
(16) Rinse in tap water, 10 min.
(17) Stain in Giemsa (see method 35)

Result

Reticulum fibres stain black.

29

Perls' Method for the Demonstration of Iron (Original method by Perls in 1867, this version by AFIP: see Luna, 1968)

Purpose

To demonstrate iron.

Fixation

10% Buffered neutral formalin, air-drying with post-fixation in 70% alcohol.

Solutions Needed

10% Potassium ferrocyanide in distilled water, 70 ml
10% Hydrochloric acid, 30 ml
(These two solutions together make up the working solution of potassium ferrocyanide–HCl.)
0.1% Kernechtrot in 5% solution of aluminium sulphate.

Procedure

(1) Bring slides to distilled water.
(2) Leave in potassium ferrocyanide solution, 5 min.
(3) Leave in working solution for 20 min.
(4) Rinse well in distilled water.
(5) Counterstain in Kernechtrot, 5 min.
(6) Rinse well in running water.
(7) Dehydrate in 95% alcohol and absolute alcohol.
(8) Clear in xylene and mount in Permount.

Result

Ferric iron bright blue.
Nuclei red.
Cytoplasm pale pink.

30

Gomori's Method for the Demonstration of Iron (Luna, 1968)

Purpose

To demonstrate iron in histiocytes, for instance in sputum.

Solutions Needed

20% Hydrochloric acid in distilled water
10% Ferrocyanide in distilled water
(These two solutions are to be mixed in equal quantities just before use.)
0.1% Kernechtrot in 5% aluminium sulphate

Fixation

10% Buffered neutral formalin or absolute alcohol.

Procedure

(1) Bring smears to distilled water.
(2) Leave in HCl–ferrocyanide solution for 30 min.
(3) Rinse thoroughly in distilled water.
(4) Counterstain in Kernechtrot if wanted.
(5) Rinse in distilled water.
(6) Dehydrate in 95% alcohol and absolute alcohol.
(7) Clear in two changes of xylene and mount.

Result

Iron pigment bright blue.
Nuclei and background red if counterstain is used.

31

Supravital Staining of Sediments in Serous Effusions (Koss, 1979, originally Foot and Holmquist, 1958)

Purpose

To differentiate histiocytes and leucocytes from neoplastic and mesothelial cells in effusions.

Solutions needed

Working solution of Neutral Red,[27] 2 ml
Working solution of Janus Green,[27] 5–10 drops
Mix the two solutions.

Procedure

(1) Place 1 or 2 drops of the dye mixture on a clean slide.
(2) Invert another slide over it.
(3) Draw the two slides apart so that an evenly distributed film is left on one side of either slide. The dye films dry quickly and can be kept indefinitely.
(4) Smear sediment on dry film and coverslip.

Result

Granules or vacuoles of histiocytes brilliant orange–red.
Polymorphonuclear leucocytes brilliant orange–red.
Mitochondria green.
Lymphocytes diffuse sky-blue.
Neoplastic or mesothelial cells are not stained.

32

Rapid Staining for Wet Sediments using either Toluidine Blue or Methylene Blue (Koss, 1979; originally Harris and Keebler, 1976)

Purpose

To stain wet sediments.

Solutions Needed

Methylene Blue[29] or Toluidine Blue[28]

Procedure

(1) Place a drop of centrifuged sediment on a glass slide.
(2) Place a drop of dye solution on the slide and mix with an applicator stick.
(3) Coverslip and leave for 2–5 min. The slide is then ready for microscopic examination. It may be temporarily preserved by applying Vaseline or hot wax around the edges.

Result

Cytoplasm and nucleus blue.

33

May-Grünwald – Giemsa Method (Romeis, 1968)

Purpose

To stain blood films, but also for general staining.

Solutions Needed

May-Grünwald's Eosin Y – Methylene Blue stain
 (stock solution 33)
Giemsa stain (working solution 30, 31 or 32)

Fixation

Air-drying.

Procedure

(1) Post-fix in May-Grünwald stock solution, by putting 20–30 drops of the solution on the slide.
(2) Add the same number of drops of distilled water, 1 min.
(3) Let fluid run off the slide.
(4) Add 10 drops of Giemsa solution, 15–20 min.
(5) Rinse well in running tap water.
(6) Blot dry.

Result

See staining result for the standardised Romanowsky – Giemsa stain (method 34)

34

The Standardised Romanowsky – Giemsa Staining Method (as recommended by the International Committee for Standardization in Haematology)

Purpose

To stain blood or bone marrow films; can also be used as general stain.

Solutions Needed

Stock solution of standardised R – G stain
Working solution of standardised R – G stain

Procedure

(1) Post-fix in stock solution for 5 min.
(2) Rinse briefly in distilled water.
(3) Stain in working solution for 25 min in the case of blood films, for 35 min in the case of bone marrow films.
(4) Rinse briefly in distilled water.
(5) Dry in air.

Result

Chromatin purple.
Nucleoli light blue.
Basophilic cytoplasm blue.
Basophilic granules purple – black.
Eosinophilic granules red – orange.
Neutrophilic granules purple.
Toxic granules black.
Platelet granules purple.
Haemoglobinised erythrocytes pink – orange.
Auer rods purple.
Doehle bodies bright blue.
Howell – Jolly bodies purple.

When used as a general stain:
Results provided separately.

35

Giemsa Staining Method (Lopez Cardozo, personal communication)

Purpose

General stain for cytological material

Solutions Needed

Giemsa stock solution, Merck No. 9204
Giemsa working solution: dilute stock solution with phosphate buffer at a ratio of 1 : 9
Sörensen's phosphate buffer at pH 6.8: mix equal quantities of $M/15$ NaH_2PO_4 and $M/15$ Na_2HPO_4; dilute the mixture with distilled water at the ratio 2 : 1 (a low molarity is needed to prevent the buffer ions from competing with the dye ions)

Fixation

Air-drying.

Procedure

(1) Post-fix in methanol for 15 min.
(2) Stain in Giemsa working solution. Staining times: in general for 20 min; material with low cell content, like urine, ascites and pleura, 10 min; material with very low cell content, like liquor and eye smears, 5 min.
(3) Rinse in tap water and leave slides to dry while they are standing upright.

Understained slides can be stained again, whereas overstained slides can be destained by leaving them in methanol for 1–3 min, followed by rinsing in tap water and air-drying.

Quick Staining with Giemsa

(1) Post-fix in methanol, 8 min.
(2) Stain in mixture of stock solution and buffer at a ratio of 1 : 1.

36

Wright's Staining Method, Lillie Modification, 1965 (Clark, 1973)

Purpose

Differentiation of blood corpuscles.

Solutions Needed

Stock stain solution: Wright's stain powder, 1 g; glycerol, 50 ml; methanol, 50 ml
Working solution: stock solution, 4 ml; acetone, 3 ml; $M/15$ phosphate buffer, pH 6.5, 2 ml; distilled water, 31 ml

Fixation

Immediately after taking sample, in methanol for 3 min.

Procedure

(1) Air-dry blood film after methanol fixation.
(2) Stain in Coplin jar with working solution for 5 min.
(3) Rinse in distilled water.
(4) Air-dry.

Result

Same as Romanowsky–Giemsa stain (method 34).

37

Gram Stain (Koss, 1979)

Purpose

To distinguish between Gram-positive and Gram-negative bacteria.

Solutions Needed

Crystal Violet[33]
Lugol's iodine[22]
1% Neutral Red in water
Acetone

Fixation

Air-drying, spray-fixation.

Procedure

(1) Bring slide to distilled water.
(2) Stain in Crystal Violet, 1–2 min.
(3) Dip in tap water, 10 dips.
(4) Stand in Lugol's iodine, $\frac{1}{2}$–2 min.
(5) Dip in tap water, 10 dips.
(6) Dip in acetone, 10–15 dips.
(7) Rinse in running tap water 5–10 min.
(8) Stain in Neutral Red, $\frac{1}{2}$–1 min.
(9) Dip in two changes of tap water, 15 dips each.
(10) Dip in two changes of 95% alcohol, 15 dips each.
(11) Dip in absolute alcohol, 15 dips each.
(12) Dip in two changes of xylene, 15 dips each.
(13) Mount.

Result

Gram-positive bacteria blue–black.
Gram-negative bacteria red.
Background pink.

38

Ferrous Iron Uptake for Melanin (Koss, 1979)

Purpose

To stain melanin.

Solutions Needed

2.5% Ferrous sulphate in water
1.0% Potassium ferricyanide in 1% acetic acid
0.1% Kernechtrot in 5% ammonium sulphate
(dissolved with heating)
1% Acetic acid

Fixation

10% Buffered formaldehyde or any routine fixative.

Procedure

(1) Bring slides to distilled water.
(2) Leave in ferrous sulphate, 1 hour.
(3) Rinse in 4 changes of distilled water, 1 min each.
(4) Leave in potassium ferricyanide, 30 min.
(5) Dip in 1% acetic acid, 15 dips.
(6) Dip in distilled water, 15 dips.
(7) Stain in Kernechtrot, 1–2 min.
(8) Dip in distilled water, 15 dips.
(9) Dip in 96% alcohol, 15 dips.
(10) Dip in 100% alcohol, 15 dips.
(11) Clear in xylene, 15 dips.
(12) Mount.

Result

Melanin dark blue to green.
Background red to pink.

39

40

Destaining of Papanicolaou Stain (Koss, 1979)

Purpose

To remove Papanicolaou stain.

Procedure

If slides are already coverslipped, remove coverslip by soaking in xylene.
(1) Rinse slides in 2 to 3 rinses each of absolute alcohol, 95% alcohol and water. This removes counterstain.
(2) Leave slides in aqueous 0.2–0.5% hydrochloric acid, 5 min to 1 hour to remove dyes. Check under microscope to see whether all dye is removed.
(3) Remove acid by rinsing in running tap water, 10–15 min. To make sure all acid is removed, place slides in Scott's tap water substitute with pH of about 8.2 and rinse again in tap water.

Result

This process leaves the slides ready either for restaining with Papanicolaou's stain in case the first stain was poor or with a particular stain to colour specific cell components.

Note: Destaining in general: Leave the slides in the dye solvent until colour has disappeared.

Destaining of Romanowsky – Giemsa Stain

Purpose

To remove stain after poor staining.

Procedure

Leave slides in methanol until colour has disappeared. NB. The methanol should not contain any acid. If it does, subsequent staining with R–G stain will yield not purple but blue nuclei.

Appendix 3

Staining Solutions

1

Delafield's Haematoxylin (Original Recipe, taken from Bolles-Lee's *Microtomist's Vademecum*)

Purpose

To stain nuclei regressively.

Chemicals Needed

Haematoxylin, 3.2 g
95% Alcohol, 20 ml
Saturated aqueous aluminium alum, 320 ml
Glycerine, 80 ml
Methanol, 80 ml

Preparation

(1) Dissolve Haematoxylin in alcohol.
(2) Add drop by drop to ammonium alum.
(3) Leave exposed to air and light for 1 week.
(4) Filter.
(5) Add glycerine and methanol.
(6) Leave for 6–8 weeks till solution becomes dark.
(7) Before using, dilute with an equal amount of distilled water.

The solution keeps very well.
Differentiation in acid and blueing in alkali are needed.

Result

Dark-blue nuclei.

2

Harris's Haematoxylin (Romeis, 1968)

Purpose

To stain nuclei regressively.

Chemicals Needed

Haematoxylin, 5 g
Absolute alcohol, 50 ml
Potassium or ammonium alum, 100 g
Mercuric oxide, 2.5 g
Glacial acetic acid, 40 ml or none
Distilled water, 950 ml

Preparation

(1) Dissolve Haematoxylin in alcohol.
(2) Dissolve alum in water by heating.
(3) Mix Haematoxylin and alum solution and heat mixture to boiling point.
(4) Remove from heat; add mercuric oxide.
(5) Cool quickly by plunging in cold water as soon as mixture turns dark purple.
(6) Filter.

The stain solution can be used immediately and keeps for a long time.
Differentiation in acid and blueing in alkali are needed.

Result

Dark-blue nuclei.

3

Mayer's Haematoxylin

Purpose

To stain nuclei progressively.

Chemicals Needed

Haematoxylin, 1 g
Distilled water, 1000 ml
Sodium iodate, 0.2 g
Aluminium alum, 50 g
Glacial acetic acid, 40 ml or none
    or citric acid, 1 g
Chloral hydrate, 50 g

Preparation

(1) Combine all ingredients in order of listing.
(2) Stir on magnetic mixer for about an hour at
    room temperature.

The solution can be used immediately and keeps
well.
Differentiation in acid should not be necessary;
blueing is.

Result

Blue nuclei.

4

Weigert's Haematoxylin (Romeis, 1968)

Purpose

To stain nuclei regressively.

Chemicals Needed

Haematoxylin, 5 g
95% Alcohol, 500 ml
Ferric chloride, 5.8 g
Concentrated HCl, 5 ml
Distilled water, 490 ml

Preparation

(1) Dissolve Haematoxylin in alcohol.
(2) Dissolve ferric chloride in distilled water and
    add HCl.
(3) Mix the two solutions in equal quantities just
    before staining.

The two solutions should be kept separate. In that
way they can be kept for years. The mixture can
only be kept for 8 days.
Differentiation in acid and blueing in alkali are not
needed.

Result

Black nuclei.

5

Cole's Haematoxylin (Cole, 1943)

Purpose

To stain nuclei progressively.

Chemicals Needed

Haematoxylin, 0.5 g
Iodine, 0.5 g
Warm water, 250 ml
96% Alcohol, 50 ml
Saturated ammonium alum solution in water, 700 ml

Preparation

(1) Dissolve Haematoxylin in warm water.
(2) Add the iodine solution.
(3) Add ammonium alum.
(4) Bring to the boil.
(5) Allow to cool quickly.

For a stronger stain the dye quantity may be doubled.
The solution keeps very well.
Check pH. It should not be higher than 3.0 for optimal results.

Result

Blue nuclei.

6

Gill's Haematoxylin (Original Recipe)

Purpose

To stain nuclei progressively.

Chemicals Needed

Haematoxylin, 2.0 g
Sodium iodate, 0.2 g (The amount of sodium iodate should be very accurately measured (accuracy $\pm$ 0.001 g.)
Aluminium sulphate, $Al_2(SO_4)_3.18H_2O$, 17.6 g
Ethylene glycol, 250 ml
Distilled water, 730 ml
Glacial acetic acid, 20 ml

Preparation

(1) Mix all ingredients in order of listing.
(2) Stir for 1 h on magnetic mixer at room temperature.

The solution keeps very well.
Differentiation in weak acid and blueing in Scott's tap water substitute needed.

Result

Dark-blue nuclei.
Nucleoli less darkly stained.

7

Carminic Acid, Carmalum (Originally Mayer, Taken from Romeis, 1968)

Purpose

To stain nuclei.

Chemicals Needed

Carminic acid, 1.0 g
Potassium alum, 10.0 g
Distilled water, 200.0 ml

Preparation

(1) Dissolve dye and potassium alum in water with heat.
(2) Leave to cool.
(3) Filter.
(4) Add 1 ml formol or 0.2 g salicylic acid to prevent mould.

8

Coelestine Blue B (Yasumatsu, 1977)

Purpose

Nuclear stain in all procedures where Haematoxylin is used.

Chemicals Needed

Coelestine Blue B, 1.0 g
4 N HNO_3, 0.4 ml
5% Ferric alum, 200.0 ml at 60°C

Preparation

(1) Add HNO_3 to Coelestine Blue in beaker, stir and mix with glass rod.
(2) Add Fe–alum and mix well till dissolved.
(3) Filter when mixture has cooled down to room temperature.
(4) Filter before use.

9

Coelestine Blue B (Gray *et al.*, 1956)

Purpose

Nuclear stain in all procedures where Haematoxylin is used.

Chemicals Needed

Coelestine Blue B, 1.0 g
Concentrated H_2SO_4, 0.5 ml
Ferric alum, 2.5 g
Distilled water, 100.0 ml
Glycerol, 14.0 ml

Preparation

(1) Rub together dye and sulphuric acid in beaker.
(2) Mix ferric alum, distilled water and glycerol at 50°C.
(3) Mix the two solutions.
(4) Leave to cool.
(5) Adjust to pH 0.8 with concentrated sulphuric acid.

10

Gallocyanin Chrome Alum (Sandritter *et al.*, 1966; also Brown and Scholtz, 1979)

Purpose

Nuclear stain for the quantitative assessment of nucleic acids and as a general nuclear stain to replace Haematoxylin.

Chemicals Needed

Chrome alum, 5 g
Gallocyanin, 0.15 g
Distilled water, 100 ml
HCl or NaOH

Preparation

(1) Dissolve chrome alum in water.
(2) Add dye and heat slowly until dissolved.
(3) Bring the mixture to pH 1.64 with a few drops of concentrated HCl, checking carefully with a pH meter.
(4) The dye solution will only keep for about 3 weeks.

NB The stain can be used in the same sequences as Haematoxylin. It has the advantage of being a progressive stain and the disadvantage of needing a very long staining time.

11

Methyl Green–Pyronin Y (Taft, 1951; also Clark, 1973)

Purpose

To stain DNA and RNA differentially.

Chemicals Needed

Methyl Green, 0.5 g
Pyronin Y (pure dye), 0.05 g
Hot distilled water, 100 ml
Chloroform

Preparation

(1) Add Methyl Green to hot water.
(2) Cool.
(3) When cool, extract the solution in separating funnel with successive 20–30 ml aliquots of chloroform until the latter remains colourless or is only slightly tinged with green.
(4) Add Pyronin Y and shake to dissolve.
(5) Store in amber, glass-stoppered bottle.

No filtering needed before use. The solution can be reused.

12

Stains and Solutions for Love's Toluidine Blue Molybdate Method for the Staining of Nucleolini (Love *et al.*, 1973)

Solutions Needed

McIlvaine's buffer pH 3.0: M/10 citric acid, 16.5 ml; M/5 disodium phosphate, 3.5 ml. These two stock solutions are made in 25% methanol. The actual buffer is made up by adding water at a ratio of 1 : 25
Staining solution: Toluidine Blue, 1 g; distilled water, 100 ml. Leave the solution overnight in a shaking machine. Working solution: stock Toluidine Blue solution, 0.75 ml; McIlvaine's buffer, 49.25 ml.
Lugol's iodine (see Stain 22)
Sodium thiosulphate: $Na_2S_2O_3.5H_2O$, 5.0 g; distilled water, 100.0 ml.
Deoxyribonuclease solution: Trishydroxymethyl aminomethane buffer (0.045 M $MgCl_2.6H_2O$ and 5 mM $CaCl_2$ per 100 ml distilled water: the buffer is brought to pH 7.3 by adding 1N HCl); DNA-ase, 1 mg. Dissolve the DNA-ase in the buffer. Use the solution at 37°C
Ammonium molybdate: $3(NH_4)_2$ 0.7 $MoO_3.4H_2O$, 15.0 g; demineralised water, 100.0 ml (the water should be demineralised double glass or quartz distilled).
Tertiary butanol: to be used at a temperature above its melting point (= 22.5°C) or mixed with a small amount of 95% alcohol.

13

Schiff's Reagent

Purpose

To be used in the periodic acid Schiff reaction for the demonstration of glycogen, and in the Feulgen reaction for the demonstration of DNA.

Chemicals Needed

Basic Fuchsin, 1 g
$NaHSO_3$, 2 g
1N HCl, 20 ml
Charcoal, 500 mg
Distilled water, 80 ml

Preparation

(1) Dissolve the dye in the water.
(2) Add $NaHSO_3$ and HCl.
(3) Stopper tightly and shake at intervals for 2 h. Solution should now be clear and light yellow.
(4) Add charcoal, shake 1 min and filter. Solution should be clear and colourless.

Keep the solution at 5°C. It keeps for several months at this temperature. If the solution turns pink, it should be discarded.

14

Leiden – EA (Drijver and Boon, 1983*b*)

Purpose

Cytoplasmic stain in Papanicolaou sequence in which acidity is variable.

Chemicals Needed

Eosin Y, 1.8 g
Light Green Y, 0.7 g
50% Alcohol, 1000 ml
for pH 4.6, 3.0 g Phosphotungstic acid

The pH can be adjusted as desired.
When OG is used, the pH of the EA has to be 6.5. This can be achieved by adding lithium carbonate.

Preparation

Mix all ingredients.
Check pH.

15

Eosin–Azure (EA) in Papanicolaou's Staining Method (Koss, 1979, slightly modified)

Purpose

To stain cytoplasm differentially in the Papanicolaou sequence.

Chemicals Needed

	EA 35	EA 65	EA for Saccomano's method
Eosin Y	2.25 g	2.25 g	2.25 g
Light Green Y	0.45 g	0.23 g	0.33 g
Phosphotungstic acid	2.0 g	6.0 g	—
Saturated lithium carbonate	10 drops	—	—
95% Alcohol			1000 ml

Koss recommends making aqueous stock solutions first. The quantities given here are calculated from Koss's table for the preparation of stock solutions. It is, however, quite possible to make up the solutions from the quantities given above, and dissolve immediately in 96% alcohol.

We have omitted Bismarck Brown since this does not add to the staining effect. It is also omitted in most commercial EA stains.

16

Gill's Modified EA (Gill, 1977)

Purpose

General cytoplasmic stain in Papanicolaou's staining sequence.

Chemicals Needed

Eosin Y in water, 4.0 g
Light Green Y in water, 0.3 g
95% Alcohol, 700 ml
Methanol, 250 ml
Glacial acetic acid, 20 ml
Phosphotungstic acid, 2 g

Preparation

Mix all ingredients.

17

Gill's Modified Orange G-6 (Gill, 1977)

Purpose

To stain highly keratinised cells in the Papanicolaou staining sequence.

Chemicals Needed

4.0 g Orange G in 95% alcohol, 20 g
95% Alcohol, 980 ml
Phosphotungstic acid, 0.15 g

Preparation

Mix all ingredients.

18

Orange G (Koss, 1979)

Purpose

To stain highly keratinised cells in the Papanicolaou staining sequence

Chemicals Needed

Orange G, 5 g
96% Alcohol, 1000 ml
Phosphotungstic acid, 0.15 g

Preparation

Mix all ingredients.

19

Orange G – Leiden (unpublished)

Purpose

To stain highly keratinised cells differentially in Papanicolaou's staining method.

Chemicals Needed

Orange G, 2.0 g
Phosphotungstic acid, sufficient to give a pH between 2.0 and 2.8 for optimal staining
50% Alcohol, 1000 ml

Preparation

Mix all ingredients. Check pH. The pH of the EA should be around 6.5.

20

Shorr's Stain (Shorr, 1942)

Purpose

For hormonal assessment.

Chemicals Needed

Biebrich Scarlet, 0.5 g
Orange G, 0.25 g
Fast Green FCF, 0.075 g
Phosphotungstic acid, 0.5 g
Phosphomolybdic acid, 0.5 g
Glacial acetic acid, 1.0 ml

Preparation

Mix all ingredients.

21

Rakoff's Stain (Rakoff, 1960)

Purpose

To stain gynaecological smears for hormonal assessment.

Chemicals Needed

5% Light Green Y in water, 83 ml
1% Eosin Y in water, 17 ml

Preparation

Mix the two solutions.

22

Lugol's Iodine (Luna, 1968)

Purpose

To stain glycogen; also for the removal of sublimate.

Chemicals Needed

Iodine, 1.0 g
Potassium iodide, 2.0 g
Distilled water, 100.0 ml

Preparation

Mix all ingredients.

23

Best's Carmine (Luna, 1968)

Purpose

To stain glycogen.

Chemicals Needed

For stock solution:
  Carmine, 2.0 g
  Potassium carbonate, 1.0 g
  Potassium chloride, 5.0 g
  Distilled water, 60.0 ml
  28% NH$_4$OH, 20.0 ml
For working solution:
  Carmine stock solution, 10.0 ml
  28% NH$_4$OH, 15.0 ml
  Methanol, 15.0 ml

Preparation

Stock solution:

(1) Boil together first four ingredients in an evaporating dish very gently for several minutes.
(2) When cool add ammonium hydroxide. Store in a refrigerator.

Working solution: mix all ingredients.

24

Oil Red O (Luna, 1968)

Purpose

To demonstrate fat in cytoplasm.

Chemicals Needed

Oil Red O, 0.5 g
Propylene glycol, 100% 100.0 ml

Preparation

(1) Add a small amount of propylene glycol to the dye and mix well; crush larger pieces.
(2) Gradually add the remainder of the propylene glycol stirring occasionally.
(3) Heat gently until solution reaches 95°C, do not allow temperature to go above 100°C. Stir while heating.
(4) Filter through coarse filter paper while still warm.
(5) Allow to stand overnight at room temperature.
(6) Filter through Seitz filter with the aid of a vacuum. (When using Seitz filter put rough surface of filter uppermost.) If solution becomes turbid, refilter.

25

Sudan Black B (Clark, 1973)

Purpose

To demonstrate fat in cytoplasm.

Chemicals Needed

Sudan Black B, 0.7 g
Propylene glycol, 100 ml

Preparation

The preparation method is the same as that for Oil Red O.

26

Gomori's Silver Stain for Fibres (Heckner Modification)

Purpose

To stain fibres.

Chemicals Needed

10% Silver nitrate, 40 ml
Distilled water, 40 ml
10% KOH solution, 3 ml
25% Ammonia
Use acid-cleaned glassware

Preparation

(1) Add 2.5 ml of the KOH to 10 ml of the silver nitrate solution.
(2) Add 25% ammonia drop by drop while shaking the container continuously until brown precipitate is completely dissolved.
(3) Add again 4 drops of silver nitrate solution for every 10 ml of silver nitrate used.
(4) Double the volume of the solution with distilled water.

27

**Neutral Red and Janus Green (Koss, 1979;
Originally Foot and Holmquist, 1958)**

Purpose

To stain histiocytes and leucocytes differentially
from mesothelial and neoplastic cells.

Chemicals Needed

Saturated solution of Neutral Red in 100%
alcohol.
Saturated solution of Janus Green in 100%
alcohol.

Preparation of Working Solution

Neutral Red solution: 20–50 drops in 10 ml 100%
alcohol.
Janus Green solution: 15–30 drops to 10 ml 100%
alcohol.

28

**Toluidine Blue (Koss, 1979; Originally Harris and
Keebler, 1976)**

Purpose

To stain quickly for wet sediment examination.

Chemicals Needed

Toluidine Blue, 0.5 g
95% Alcohol, 20.0 ml
Distilled Water, 80 ml

Preparation

(1) Dissolve dye in alcohol and add water.
(2) Filter and store in a dark bottle in a refrigerator.

29

30

Methylene Blue (Koss, 1979; Originally Harris and Keebler, 1976)

Giemsa Stain (Original Recipe)

Purpose

To use as a quick stain for wet sediment examination.

Chemicals Needed

Methylene Blue, 1.5 g
95% Alcohol, 30.0 ml
0.1 N Potassium hydroxide, 2 ml

Preparation

(1) Dissolve the dye in the alcohol and add the KOH.
(2) Store in a dark bottle in a refrigerator.

Purpose

Differentiation of blood corpuscles, general stain.

Chemicals Needed

Azure II – Eosin Y, 3.0 g (Azure II is a mixture of Azure B and Methylene Blue in equal quantities)
Azure II, 0.8 g
Glycerol: for blood, 250 g or 200 ml; for tissue, 125 g or 100 ml
Methanol (neutral, acetone free): for blood, 250 g or 312 ml; for tissue, 375 g or 457 ml

Preparation of Stock Solution

(1) Mix glycerol and methanol.
(2) Dissolve the dyes in the mixture.
(3) Stand at room temperature overnight.
(4) Shake well for 5–10 min.
(5) Pour without filtering into dark screw-cap bottle.
(6) Store at room temperature.

Preparation of Working Solution

Mix 5 ml of stock solution with 65 ml of water.

31

Giemsa Stain (Lillie Modification, 1943b)

Purpose

To stain blood corpuscles and to use as general stain.

Chemicals Needed

Azure A – Eosinate, 0.5 g
Azure B – Eosinate, 2.5 g
Methylene Blue – Eosinate, 2.0 g
Methylene Blue chloride, 1.0 g
Glycerol, 375.0 ml
Methanol, 375.0 ml

Preparation of Stock Solution

(1) Mix glycerol and methanol.
(2) Dissolve dyes in the mixture.
(3) Stand at room temperature overnight.
(4) Shake well for 5–10 min.
(5) Pour into dark screw-cap bottle without filtering.
(6) Store at room temperature.

Preparation of Working Solution

Mix 5 ml of stock solution with 65 ml of water or phosphate buffer, pH 6.8.

32

Standardised Romanowsky – Giemsa Stain (Recommended by the International Committee for Standardization in Haematology)

Purpose

Differentiation of blood corpuscles; can also be used as general stain.

Chemicals Needed

For stock solution
    Azure B – thiocyanate, 3 g
    Eosin – disodium salt, 1 g
    Methanol, 600 ml
    Dimethylsulphoxide, 400 ml
For working solution
    Stock solution
    Hepes buffer, pH 6.8

Preparation

Stock solution:
(1) Dissolve Azure B: add 400 ml of DMSO to the Azure B powder at 37°C.
(2) Add 1 g of Eosin Y to 600 ml methanol. Wait until the dye is completely dissolved.
(3) Slowly mix Azure B – DMSO with Eosin – methanol.
This solution keeps for many months at room temperature if stored in a glass bottle in the dark. To improve stability the stock solution may be kept at lower pH (4.0) by adding hydrochloric acid.
Working solution: Mix stock solution and buffer at a ratio of 1 : 15. For rapid staining, mixtures with a higher proportion of stock solution may be used.

33

34

May-Grünwald's Eosin Y – Methylene Blue Stain (from Clark's *Staining Procedures*, 1981)

Purpose

To stain blood films and for general staining.

Chemicals Needed

Eosin Y, 0.5 g
Methylene Blue, 0.5 g
Distilled water, 100 ml
Absolute methanol, 50 ml

Preparation of Stock Solution

(1) Mix Eosin and Methylene Blue.
(2) Filter.
(3) Dry filtrate.
(4) Wash residue and dry.
(5) Dissolve residue in methanol.

Crystal Violet in Gram Stain (Koss, 1979)

Purpose

To stain Gram-positive bacteria.

Chemicals Needed

Crystal Violet, 5.0 g
95% Alcohol, 50.0 ml
Ammonium oxalate, 2.0 g
Distilled water, 200.0 ml

Preparation

(1) Dissolve dye in alcohol.
(2) Dissolve ammonium oxalate in water.
(3) Mix the two solutions and filter before use.

Appendix 4

Dyes

This appendix provides a list of dyes either mentioned in the text or used in one of the staining methods. The following data are provided for each dye: structure (taken from *Conn's Biological Stains* (Lillie, 1977), unless otherwise stated); ionic or molecular weight; Hansch π value; conjugated bond number; net charge; absorption maximum; solubility; main uses; and the CI number. When uses for the dyes are mentioned in either appendix 2 or 3, the reader is referred to the appropriate method or stain. Further information may be found elsewhere in the book, or in *Conn's Biological Stains* (Lillie, 1977).

Ionic weight, Hansch π value, conjugated bond number and net charge are kindly provided by Dr R.W. Horobin of the University of Sheffield.

Acid Fuchsin

CI no. 42685
(Derivative of Rosanilin or Pararosanilin)

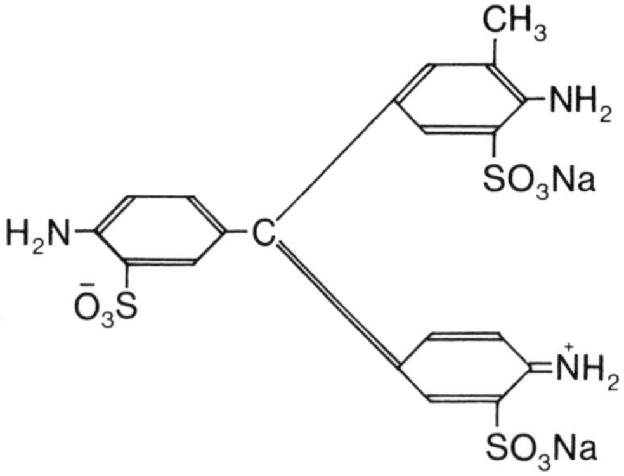

Uses

As a cytoplasmic stain in many polychrome mixtures (Mallory, 1900; Masson, 1929; Goldner, 1938; Foot, 1938; Papanicolaou, 1941).

ionic weight	540
Hansch π value	-17.3
conjugated bond number	24
charge	$2-$
absorption maximum	540–546 nm
solubility	
in water	+
in alcohol	±
Cellosolve	+

Acridine Orange

CI no. 46005

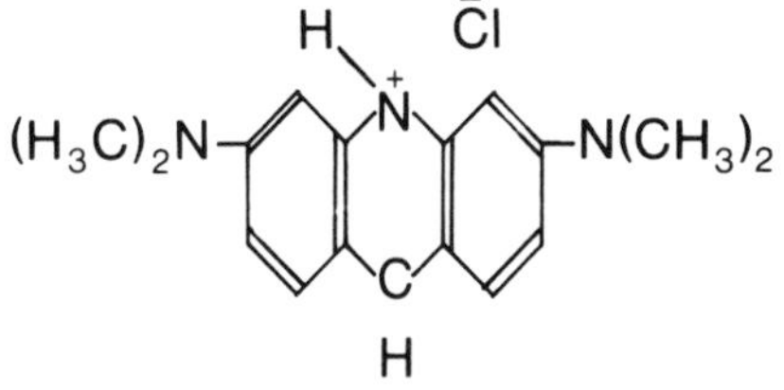

Uses

As fluorochrome for nucleic acids; as selective stain for tumour cells.

ionic weight	266
Hansch π value	-2.7
conjugated bond number	18
charge	$1+$
absorption maximum	467–497 nm
solubility	
in water	+
in alcohol	+

Alcian Blue

CI no. 74240
Structure uncertain

X = onium group, $CH_2SC \overset{NR_2}{\underset{NR_2}{\diagdown}}$

$R = C_2H_5$

Uses

As differential stain for acid polysaccharides; can be used in combination with PAS reaction.

ionic weight	1380
Hansch π value	-14.8
conjugated bond number	48
charge	4+
absorption maximum	blue

(can also stain blue–green metachromatically)

solubility	
in water	+
in alcohol	?

Alizarin Blue

CI no. 67415

$2NaHSO_3$

Uses

As a mordant nuclear dye (mordant Fe), when it stains metachromatically (Lillie, 1977).

ionic weight	291.266 without $2NaHSO_3$
Hansch π value	?
conjugated bond number	?
charge	1 −
absorption maximum	blue
solubility	
in water	±
in alcohol	−

Amido Black 10B

CI no. 20470
(Synonym Pontacyl Blue Black SX)

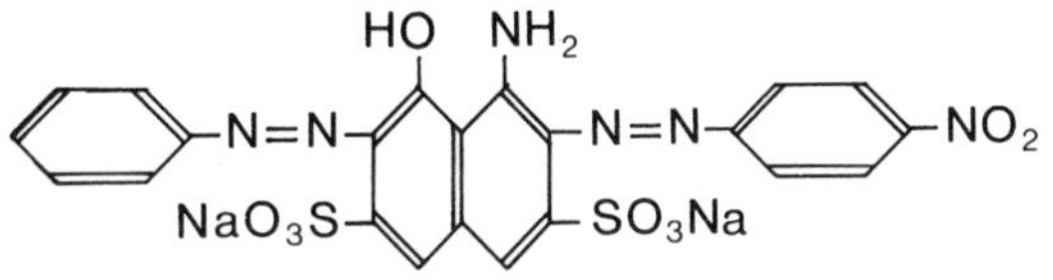

Uses

As protein stain in paper chromatography; as nucleolar stain (see Method 18). (Mundkur and Greenwood, 1968.)

ionic weight	571
Hansch π value	− 5.0
conjugated bond number	34
charge	2 −
absorption maximum	blue
solubility	
in water	+
in alcohol	+
in Cellosolve	+

Aniline Blue WS

CI no. 42755

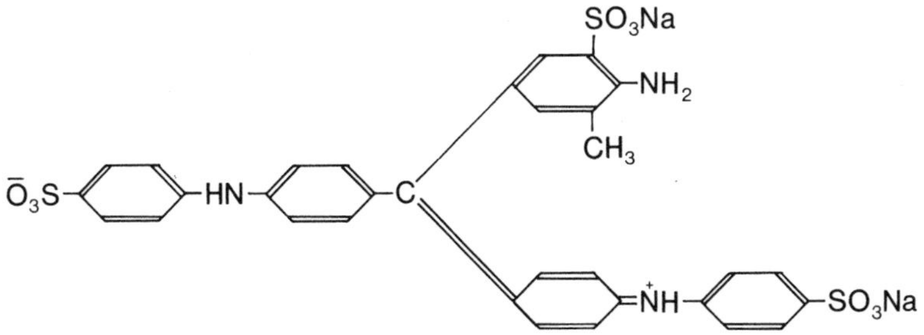

The dye is usually a mixture of a number of components, differing in degree of sulphonation and phenylation.

Uses

As a component of connective tissue stain (Mallory, 1900; Masson, 1929); in Papanicolaou's early cytoplasmic stain (Papanicolaou, 1933); in Heidenhain's Azan (Lillie, 1977).

ionic weight	692
Hansch π value	− 13.3
conjugated bond number	36
charge	2 −
absorption maximum	600 nm
solubility	
in water	+
in alcohol	−

Azure A

CI no. 52005
Structure by Comings (1975).

Uses

The dye often forms part of a Methylene Blue solution since it is a derivative of that dye. Used as a nuclear dye, and as an RNA stain in combination with a surfactant (Bennion *et al.*, 1975; see Method 20).

ionic weight	256
Hansch π value	− 4.2
conjugated bond number	18
charge	1 +
absorption maximum	630nm as orthochromatic dye
solubility	
in water	+
in alcohol	+

Azure B (= Azure I)

CI no. 52010
Structure by Comings (1975).

Uses

It was originally one of the components of Giemsa's blood stain; now it is the azure dye in the standardised blood stain (International Committee for Standardization in Haematology recommendation, 1984). See Method 34.

ionic weight	260
Hansch π value	− 3.6
conjugated bond number	18
charge	1 +
absorption maximum	650 nm

(This is the wavelength of the orthochromatic dye)

solubility	
in water	+
in alcohol	±

Azure C

CI no. 52002
Structure by Comings (1975).

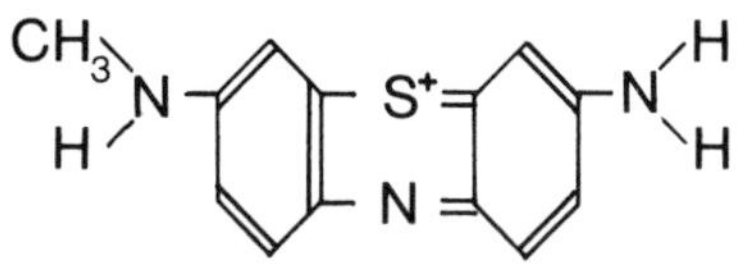

Uses

No known uses as individual dye, it is usually present in Methylene Blue solutions since it is one of the derivatives of that dye.

ionic weight	243
Hansch π value	-3.0
conjugated bond number	18
charge	$1+$
absorption maximum	620 nm

(This is the wavelength of the orthochromatic dye)

solubility
    in water $+$
    in alcohol $\pm$

Basic Fuchsin

CI no. 42500

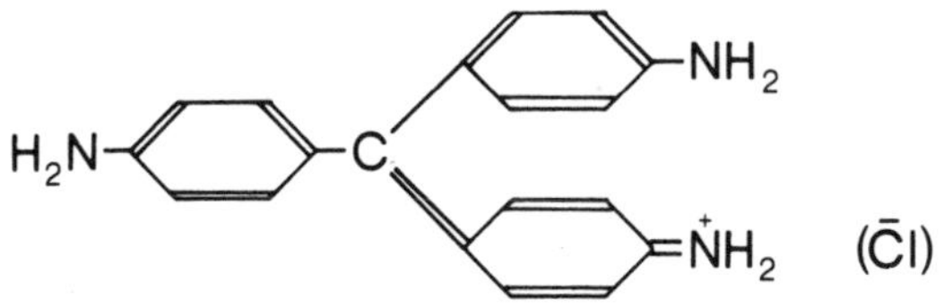

Uses

In Schiff's reagent (see Stain 13); as nuclear dye in mucin and elastic tissue staining method.

Parameters given are those for Pararosanilin since this forms the main component of the dye.

ionic weight	288
Hansch π value	-3.5
conjugated bond number	24
charge	$1+$
absorption maximum	545 nm

solubility
    in water $+$
    in alcohol $-$

Biebrich Scarlet

CI no. 26905

Uses

As cytoplasm stain in Shorr's polychrome stain for hormonal assessment (see Method 6). For staining basic proteins at graded pH (Spicer, 1961). For staining Barr bodies in Guard's method (see Method 21).

ionic weight	510
Hansch π value	-3.5
conjugated bond number	30
charge	$2-$
absorption maximum	503 nm
solubility	
in water	+
in alcohol	−
in Cellosolve	−

Bismarck Brown Y

CI no. 21000

Uses

For vital staining; as component in Papanicolaou's EA stain.

ionic weight	349
Hansch π value	-5.7
conjugated bond number	27
charge	$1+$
absorption maximum	?
solubility	
in water	+
in alcohol	±

Brilliant Cresyl Blue

CI no. 51010

Uses

As vital stain for the staining of acid mucopolysaccharides.

molecular weight	332
Hansch π value	?
conjugated bond number	?
charge	$1+$
absorption maximum	630
solubility	
in water	+
in alcohol	+

Carmine and Carminic Acid

CI no. 75470
The dye principle of Carmine is Carminic Acid.
There are other possibilities for the structure.

Uses

As chromalum carmine (Fyg's: Lillie, 1977) giving
blue–black nuclei; in Best's carmine for glycogen
(see Method 22).

ionic weight	492
Hansch π value	− 2.02
conjugated bond number	24
charge	1 −
absorption maximum	red in Best's
solubility	
in water	±
in alcohol	±

Chlorazol Black E

CI no. 30235

Uses

As direct nuclear and cytoplasmic dye, very good
for photography.

ionic weight	735
Hansch π value	− 3.4
conjugated bond number	49
charge	2 −
absorption maximum	black
solubility	
in water	+
in alcohol	±
in other organic solvents	−
in Cellosolve	±

Coelestine Blue B

CI no. 51050

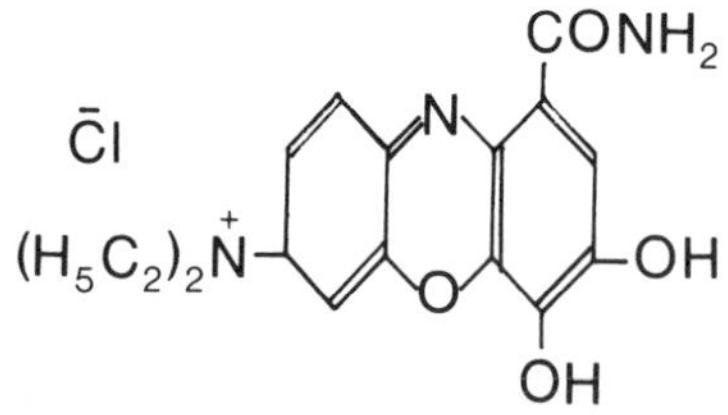

Uses

As nuclear stain, as substitute for Haematoxylin.

ionic weight	328
Hansch π value	-5.9
conjugated bond number	21
charge	$1+$
absorption maximum	654 nm
solubility	
in water	$+$
in alcohol	$+$

Coomassie Brilliant Blue

CI no. 42655
Structure by Tas *et al.* (1980).

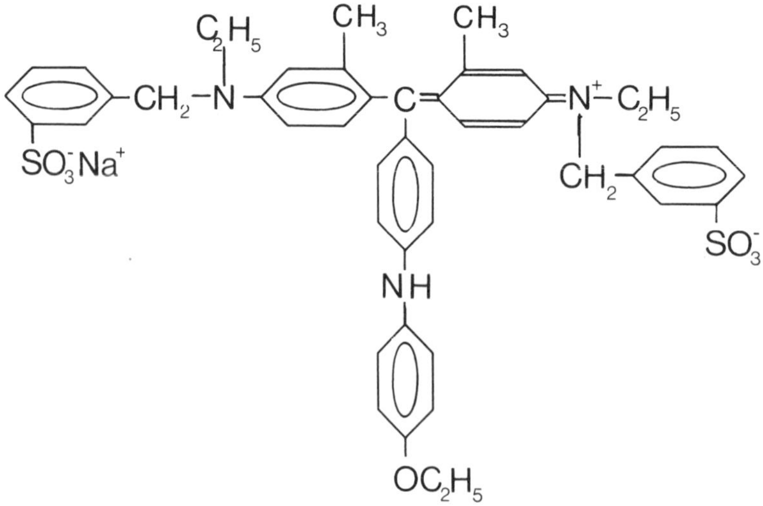

Uses

As protein stain in quantitative cytochemistry (Tas *et al.*, 1980).

molecular weight	818
Hansch π value	-1.38
conjugated bond number	43
charge	$1-$
absorption maximum	590 nm

(This is the wavelength when the dye is bound to protein, in solution the colour is dependent on pH.)

solubility	
in water	$\pm$
in alcohol	$+$

Cresyl Violet

No CI number.

The formula is that of Cresyl Acetate, which is one of the forms in which the dye is used. However the structure is undecided. No data on parameters are available.

Crystal Violet (synonym Gentian Violet)

CI no. 42555

Cuprolinic Blue (Quinolinic Phthalocyanin)

No CI number.
Structure by Tas *et al.* (1983).

Me = Methyl

Uses

As metachromatic vital stain: it stains nuclei violet and plasma blue, amyloid bodies, mast cell granules and mucin red; as protein stain at graded pH; as Nissl stain.

molecular weight	321
Hansch π value	?
conjugated bond number	?
charge	1+
absorption maximum	600 nm
solubility	
in water	+
in alcohol	+

Uses

As nuclear stain; as vital stain for amyloid bodies in fresh or fixed tissue; to stain blood platelets; in Gram stain.

ionic weight	372
Hansch π value	+ 0.1
conjugated bond number	24
charge	1+
absorption maximum	589–593 nm
solubility	
in water	±
in alcohol	+

Uses

As specific stain for nucleic acids; in combination with MgCl it is a specific stain for RNA (Tas *et al.*, 1983; see Method 19).

ionic weight	640
Hansch π value	− 21.6
conjugated bond number	44
charge	4+
absorption maximum	635 nm
solubility	
in water	+
in alcohol	?

Dinitrofluorobenzene

No CI number.
Structure by Tas *et al.* (1980).

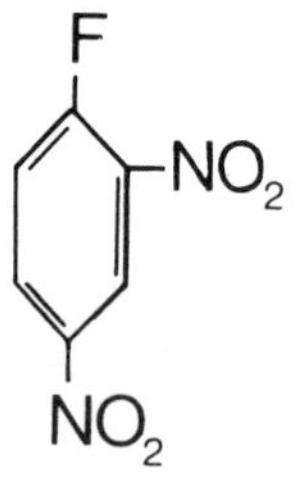

Uses

As protein stain in quantitative cytochemistry (Tas *et al.*, 1980).

ionic weight	186
Hansch π value	+ 1.78
conjugated bond number	13
charge	0
absorption maximum	400
solubility	
in water	+
in alcohol	+

Eosin Y

CI no. 45380

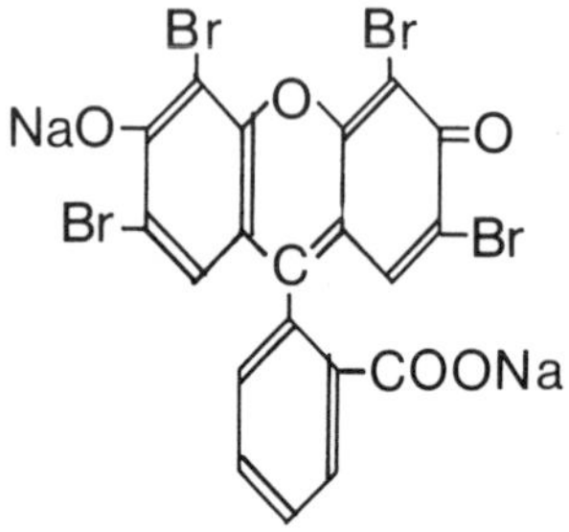

Uses

A counterstain in several polychrome stains including Papanicolaou's in Romanowsky–Giemsa stains (Methods 1–4 and 34).

ionic weight	646
Hansch π value	0.0
conjugated bond number	31
charge	2 −
absorption maximum	∼515 nm
solubility	
in water	+
in alcohol	+

Fast Green FCF

CI no. 42053

Uses

As cytoplasmic stain in Shorr's stain (Method 20, Stain 6); as protein stain for quantitative cyto-chemistry (Smetana and Busch, 1966; Tas *et al.*, 1980; Dhar and Shah, 1982).

ionic weight	763
Hansch π value	− 10.6
conjugated bond number	36
charge	2 −
absorption maximum	625 nm
solubility	
in water	+
in alcohol	+

Fluorone Black

No CI number.

Uses

As mordant nuclear stain as Haematoxylin sub-stitute, tested by Lillie *et al.* (1975*b*); it gives black nuclei.

ionic weight	320
Hansch π value	+ 3.4
conjugated bond number	27
charge	0
absorption maximum	black
solubility	
in water	+
in alcohol	±

Gallein

CI no. 45445

Uses

As nuclear stain as Haematoxylin substitute, tested by Lillie *et al.* (1975*a*); as glycogen stain (Murgatroyd and Horobin, 1969).

ionic weight	361
Hansch π value	-4.6
conjugated bond number	36
charge	$2-$
absorption maximum: stains blue with Fe as mordant	
solubility	
in hot water	$\pm$
in alcohol	$+$
in alkali	$+$
in acetone	$+$
in ether	$-$
in chloroform	$-$
in benzene	$-$

Gallo Blue E

CI no. 51040

Uses

As nuclear dye as Haematoxylin substitute, tested by Lillie *et al.* (1976*b*).

No data available.

Gallocyanin

CI no. 51030

Uses

As chrome alum mordant dye; as a substitute for Haematoxylin; for quantitative spectrophotometry.

ionic weight	301
Hansch π value	-8.0
conjugated bond number	22
charge	$1+$
absorption maximum	$\pm\,636$ nm
solubility	
in water	$-$
in alcohol	$+$
in Cellosolve	$+$
in glycol	$+$

Structure by Horobin (1982).

Haematein

CI no. 75290

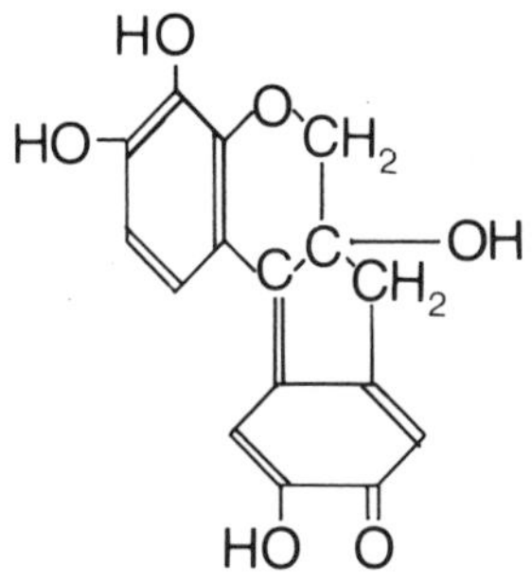

Uses

Since it forms the actual dye in a Haematoxylin solution, it is used as a nuclear dye.

ionic weight	300
Hansch π value	+ 1.6
conjugated bond number	18
charge	0
absorption maximum	445 nm
solubility	
in water	+
in alcohol	+
in Cellosolve	+
in glycol	+

Haematoxylin

CI no. 75290

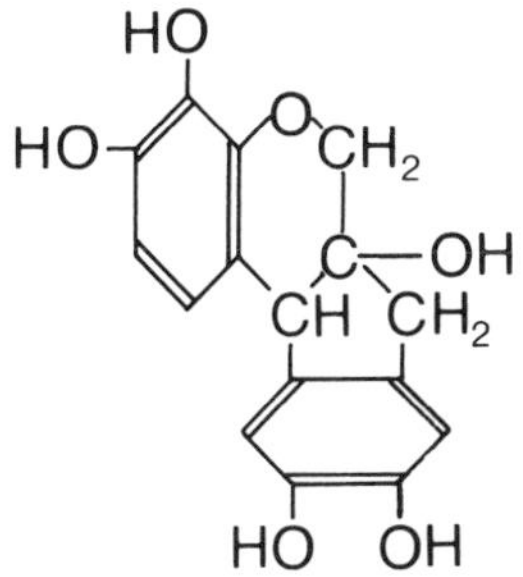

Uses

Mainly as nuclear dye after oxidation to Haematein; Lillie (1977) mentioned three more main uses.

ionic weight	302
Hansch π value	+ 1.0
conjugated bond number	16
charge	0
absorption maximum	292 nm
solubility	
in water	+
in alcohol	+
in Cellosolve	+
in glycol	+

Janus Green B

CI no. 11050

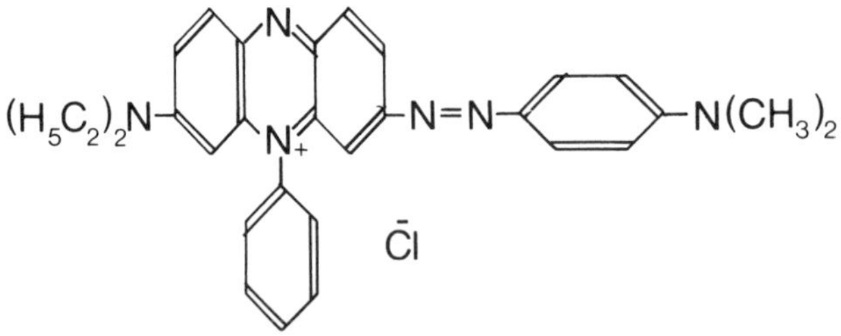

$(H_5C_2)_2N$—...—$N=N$—...—$N(CH_3)_2$

$\bar{C}l$

Uses

As supravital stain (see Method 31).

ionic weight	300
Hansch π value	+ 1.6
conjugated bond number	18
charge	0
absorption maximum	400 and
	610–623 nm
solubility	
in water	+
in alcohol	±

Kernechtrot

CI no. 60760

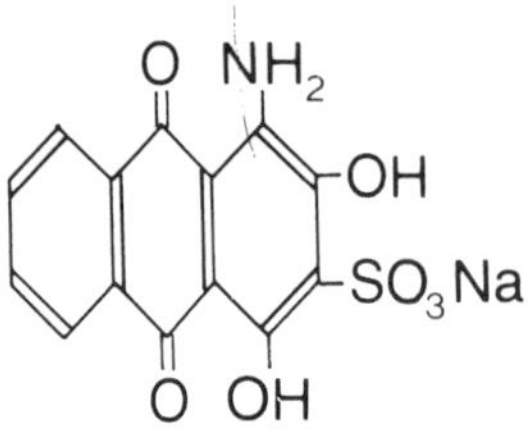

O NH$_2$
—OH
—SO$_3$Na
O OH

Uses

For the demonstration of calcium; also for the demonstration of iron, as in Perls' and Gomori's methods (see Methods 29 and 30).

molecular weight	357
Hansch π value	?
conjugated bond number	?
charge	1 –
absorption maximum	red
solubility	
in water	±
in alcohol	?

Light Green Y

CI no. 42095

Uses

As cytoplasmic stain in many polychrome stains, including Papanicolaou's (see Stains 14–16); as protein stain for quantitative cytochemistry (Oud *et al.*, 1984).

ionic weight	747
Hansch π value	− 9.5
conjugated bond number	34
charge	2 −
absorption maximum	428 and 629–634 nm
solubility	
in water	+
in alcohol	−
in Cellosolve	±

Luxol Fast Blue G

No CI number.

Uses

To stain myelin sheaths (polar lipids).

No data available on parameters.

Solubility	
in water	−
in alcohol	+
in Cellosolve	+
in methanol	+
in isopropanol	+
in acetone	+

Methyl Green

CI no. 42585

Uses

As differential stain for DNA in combination with Pyronin Y which stains RNA (see Methods 14 and 15).

ionic weight	387
Hansch π value	-5.5
conjugated bond number	23
charge	$2+$
absorption maximum	420 and
	630–634 nm
solubility	
in water	$+$
in alcohol	$-$

Methylene Blue

CI no. 52015
Structure by Comings (1975).

Uses

As direct nuclear dye; as vital stain for wet sediments; in the early blood stains where it was thought to be the main nuclear dye.

ionic weight	284
Hansch π value	-2.7
conjugated bond number	18
charge	$1+$
absorption maximum	660
solubility	
in water	$+$
in alcohol	$+$
in Cellosolve	$+$
in glycol	$+$

Methylene Violet Bernthsen

CI no. 52041

$(H_3C)_2N$—

Uses

As nuclear stain in MacNeal's blood stain (Lillie, 1977).

ionic weight	256
Hansch π value	+ 2.9
conjugated bond number	18
charge	0
absorption maximum	+ 580 nm

(This is the wavelength of the orthochromatic dye.)

solubility	
in water	−
in alcohol	+
in ether	+
in chloroform	+

Naphthol Yellow S

CI no. 10316

Structure by Tas *et al.* (1980).

Uses

As protein stain for quantitative cytochemistry can be used in sequence with the Feulgen reaction; Method 13 (Tas *et al.*, 1980).

ionic weight	312
Hansch π value	− 5.97
conjugated bond number	18
charge	2 −
absorption maximum	± 420 nm
solubility	
in water	+
in alcohol	±

Neutral Red

CI no. 50040

There are two other possibilities for the structure. This is the orthoquinoid structure given by the Colour Index in 1956.

Uses

As vital stain in combination with Janus Green.

ionic weight	?
Hansch π value	?
conjugated bond number	?
charge	?
absorption maximum	530 nm
solubility	
in water	+
in alcohol	±

Nile Blue

CI no. 51180

Uses

As fat stain; the sulphate differentiates between neutral fats and fatty acids (see Clark, 1973).

ionic weight	318
Hansch π value	-2.4
conjugated bond number	23
charge	$1+$
absorption maximum	± 640 nm
solubility	
in hot water	$+$
in alcohol	$+$

Oil Red O

CI no. 26125

Uses

As fat stain (see Method 27).

ionic weight	409
Hansch π value	$+8.99$
conjugated bond number	30
charge	0
absorption maximum	± 550 nm
solubility	
in water	$-$
in alcohol	$+$
in acetone	$+$
in isopropanol	$+$

Orange G

CI no. 16230

Uses

As a cytoplasmic stain in many polychrome stains, especially to demonstrate keratin in Papanicolaou's sequence; as a general protein stain in quantitative cytochemistry (Oud *et al.*, 1984).

ionic weight	406
Hansch π value	-5.2
conjugated bond number	21
charge	$2-$
absorption maximum	± 480 nm
solubility	
in water	$+$
in alcohol	$\pm$
in Cellosolve	$+$
in glycol	$+$

Orcein

CI no. (Ed. 1) 1242
No specific structure can be given for this since
several dye fractions may be present. No characteristic parameters are available.

Uses

It is used as a nuclear dye (it colours chromatin
bright red).

Pararosanilin

CI no. 42500

Uses

Together with Rosanilin it forms the main constituent of Basic Fuchsin; hence it is used for the
same purposes.

ionic weight	288
Hansch π value	-3.5
conjugated bond number	24
charge	$1+$
absorption maximum	± 545 nm
solubility	
in water	$\pm$
in alcohol	$+$

Picric acid

CI no. 10305

Uses

As a fixative in Bouin's; in combination with other
anionic dyes for differentiation after dyeing with
Haematoxylin.

ionic weight	228
Hansch π value	-2.8
conjugated bond number	17
charge	$1-$
absorption maximum	± 360 nm
	in alcohol
solubility	
in water	$\pm$
in alcohol	$+$
in benzene	$+$
in chloroform	$\pm$
in ether	$\pm$

Ponceau de Xylidine

CI no. 16150

Uses

As component of Masson's trichrome stain; as cytoplasmic stain.

ionic weight	434
Hansch π value	-4.0
conjugated bond number	21
charge	$2-$
absorption maximum	± 500 nm
solubility	
in water	$+$
in alcohol	$-$
in acetone	$-$

Pontacyl Dark Green B

CI no. 20495

Uses

As nucleolar dye (Bedrick, 1970) (see Method 16).

ionic weight	541
Hansch π value	-5.63
conjugated bond number	32
charge	$2-$
absorption maximum	bluish green
solubility	
in water	$+$
in alcohol	$\pm$
in Cellosolve	$+$

Pyronin Y

CI no. 45005

Rhodamine B

CI no. 45170

Rhodanile Blue

No CI number.

Uses

A combination of Rhodamine B and Nile Blue; used as nuclear dye, as Haematoxylin substitute (Gurr, 1975).

Data on all parameters not available.

Uses

As specific dye for RNA in combination with Methyl Green which stains DNA; Methods 14 and 15 (Taft, 1951; Kurnick, 1952).

ionic weight	268
Hansch π value	-2.5
conjugated bond number	18
charge	$1+$
absorption maximum	552 nm
solubility	
in water	+
in alcohol	±

Uses

As fluorescent dye; together with Nile Blue to form Rhodanile Blue.

Data not available on all parameters.

absorption maximum	bluish red, ± 556 nm
solubility	
in water	−
in alcohol	−

molecular weight	780
solubility	
in water	±
in alcohol	?

Solochrome Cyanin R

CI no. 43820
(synonyms Chromoxane Cyanin R, Mordant Blue 3)

Uses

As nuclear dye as Haematoxylin substitute (stains blue in combination with phosphoric acid: Pearse, 1957; Clark, 1979).

ionic weight	467
Hansch π value	-10
conjugated bond number	29
charge	$4-$
absorption maximum	red
solubility	
in water	+
in alcohol	?

Sudan Black B

CI no. 26150

Uses

As fat stain.

ionic weight	456
Hansch π value	$+7.4$
conjugated bond number	36
charge	0
absorption maximum	± 600 nm
solubility	
in water	$-$
in alcohol	+
in propylene glycol	+

Thionin

CI no. 52000
Structure by Comings (1975).

Uses

As nuclear stain, especially for automation (Wittekind and Hilgarth, 1979); as Haematoxylin substitute.

ionic weight	225
Hansch π value	-5.1
conjugated bond number	18
charge	$1+$
absorption maximum	± 598

(This is the wavelength of the orthochromatic dye.)

solubility	
in water	$+$
in alcohol	$+$

Toluidine Blue

CI no. 52040

Uses

As direct nuclear dye; as RNA stain in combination with a surfactant (Bennion *et al.*, 1975—see Method 20).

ionic weight	270
Hansch π value	-3.9
conjugated bond number	18
charge	$1+$
absorption maximum	± 620
solubility	
in water	$+$
in alcohol	$\pm$
in Cellosolve	$+$
in glycol	$+$

Recommended Reading

On the History of Dyeing

Baker, J.R. (1963). *Cytological Technique.* Methuen, London. Provides an excellent introduction to the mechanisms of fixation and dyeing with a wealth of historical background.

Vickerstaff, T. (1950). *The Physical Chemistry of Dyeing.* Oliver & Boyd, London.

Venkataraman, K. (1952). *The Chemistry of Synthetic Dyes.* Academic Press, New York.

On the Mechanism of Fixation, apart from Baker's

Hopwood, D. (1973). *Fixation in Histochemistry* (P.J. Stoward, Ed.) Chapman & Hall, London.

Pearse, A.G.E. (1980). *Histochemistry, Theoretical and Applied.* I. Preparative and optical technology. Churchill Livingstone, Edinburgh.

Horobin, see below.

On the Mechanism of Staining

Horobin, R.W. (1982). *Histochemistry.* Butterworths, London. The author has a very original approach to the linking mechanisms of dye and substrate, illustrated with many useful graphs and drawings.

On Cell Biology

DeRobertis, E.D.P. and DeRobertis, E.M.F. (1980). *Cell and Molecular Biology.* Holt-Saunders International editions.

On Dyes, their Classification, Application and History

Lillie, R.D. (1977). *Conn's Biological Stains.* Williams & Wilkins, Baltimore. An invaluable book with information on most individual dyes, the methods in which they are used and their history.

For Recipes of Staining Solutions and for Staining Methods

Bolles-Lee, A. (1900). *Microtomist's Vademecum.* 5th edn, Churchill, London. The author, who was an amateur, tells of his own experience and that of his contemporaries; provides very stimulating reading

Clark, G. (1981). *Staining Procedures.* Published for the Biological Staining Commission. Williams & Wilkins, Baltimore. Goes hand in hand with *Conn's Biological Stains.*

Romeis, B. (1968). *Mikroskopische Technik.* Oldenburg Verlag, Munich.

On Clinical Cytology

Koss, L.G. (1979). *Diagnostic Cytology.* Lippincott, Philadelphia.

Atlases

Lopez Cardozo, P. (1977). *Atlas of Clinical Cytology.* Lippincott, Philadelphia.

Takahashi, M. (1981). *Color Atlas of Cancer Cytology.* Thieme Verlag, Stuttgart.

References

Alfert, M. and Geschwind, J.J. (1952). Selective staining for the basic proteins of cell nuclei. *Proc. Nat. Acad. Sci.*, **39**, 991–9.

Arata, T., Sekiba, K. and Kato, K. (1978). Appraisal of self-collected cervical specimens in cytologic screening of uterine cancer. *Acta Cytologica*, **22**, 150–2.

Ascher, A.W., Turner, C.J. and De Boer, G.H. (1956). Cornification of human vaginal epithelium. *J. Anatomy*, **90**, 545–52.

Baker, J.R. (1963). *Cytological Technique*. Methuen, London.

Baker, J.R. (1970). *Principles of Biological Microtechnique*. Methuen, London.

Becker, S.N., Wong, J.Y., Marchiondo, A.A. and Davis, C.P. (1981). Scanning electron microscopy of alcohol-fixed cytopathology specimens. *Acta Cytologica*, **25**, 578–84.

Bedrick, A.E. (1970). Differential nucleolar staining of malignant and benign tissues with Pontacyl Dark Green B. *Stain Technology*, **45** (6), 273–6.

Bennion, P.J., Horobin, R.W. and Murgatroyd, L.B. (1975). The use of a basic dye (Azure A and Toluidine Blue) plus a cationic surfactant for selective staining of RNA: a technical and mechanistic study. *Stain Technology*, **50** (5), 307–13.

Bercovici, B., Diamant, Y. and Polishuk, W.Z. (1958). The use of intra-uterine device to obtain material for cytology. In: *Symposium on techniques for endometrial cytological examinations. Acta Cytologica*, **2**, 577–8.

Bernhardt, H., Gourley, R.D., Young, J.M., Shepherd, M.C. and Killian, J.J. (1961). A modified membrane-filter technique for detection of cancer cells in body fluids. *Am. J. Clin. Pathol.*, **36**, 462–4.

Berube, G.R., Powers, M.M., Kerkay, J. and Clarke, G. (1966). The gallocyanin–chrome alum stain; influence of methods of preparation on its activity and separation of active staining compound. *Stain Technology*, **41** (2), 73.

Beyer-Boon, M.E. (1977). Preparatory techniques. In: *Urinary Cytology* (H.J. de Voogt, P. Rathert and M.E. Boon, Eds) Springer Verlag, Berlin, pp. 7–14.

Beyer-Boon, M.E. and Van der Voorn-Den Hollander, M.J.A. (1978). Cell yield obtained with various cytopreparatory techniques for urinary cytology. *Acta Cytologica*, **22**, 589–93.

Beyer-Boon, M.E., Van der Voorn-Den Hollander, M.J.A., Arentz, P.W., Cornelisse, C.J., Schaberg, A. and Fox, C.H. (1979a). Effect of various routine cytopreparatory techniques on normal urothelial cells and their nuclei. *Acta Path. Microbiol. Scand., Sect. A*, **87**, 63–9.

Beyer-Boon, M.E., Arentz, P.W. and Kirk, R.S. (1979b). A comparison of thiomersal and 50% alcohol as preservatives in urinary cytology. *J. Clin. Pathol.*, **32**, 168.

Bibbo, M., Camargo, A.C.M., Valeri, V. (1969). Studies of cell lipids from the human vaginal and cervical epithelium during the menstrual cycle. *Acta Cytologica*, **13**, 260–3.

Bibbo, M., Fennessy, J.J. and Lu, C.T. (1973). Bronchial brushing technique for the cytologic diagnosis of peripheral lung lesions. A review of 693 cases. *Acta Cytologica*, **17**, 251.

Bird, C.L. (1951). *The Theory and Practice of Wool Dyeing*. Society of Dyers and Colourists, Bradford.

Boccato, P. (1981). Modification of Fischer's method for urine cytology. *Acta Cytologica*, **25**, 5–6.

Böhm, N. and Sandritter, W. (1975). DNA in human tumors. A cytophotometric study. *Curr. Top. Pathol.* **60**, 151–219.

Bolles-Lee, A. (1900). *Microtomist's Vademecum*, 5th edn. Churchill, London.

Boon, M.E. and Lykles, C. (1980). Imaginative approach to fine needle aspiration cytology. *Lancet*, 1031–2.

Boon, M.E. and Tabbers-Boumeester, M.E. (1981). *Gynaecological Cytology: A Textbook and Atlas*. Macmillan, London.

Boon, M.E., Smid, L., Van Olphen, K. and Krebber, R. (1980). An improved cytocentrifuge technique for cerebrospinal fluid cytology. *Clin. Neurol. Neurosurg.*, 205–7.

Boon, M.E., Posthuma, H.S., Ruiter, D.J. and Van Andel, J.G. (1981). Secreting peritoneal mesothelioma: Report of a case with cytological ultrastructural, morphometric and histologic studies. *Virchows Arch.* (Pathol. Anat.), **392**, 33–44.

Boon, M.E., Kwee, H.S., Alons, C.L., Morawetz, F. and Veldhuizen, R.W. (1982a). Discrimination between primary pleural and primary peritoneal mesotheliomas by morphometry and analysis of the vacuolization pattern of the exfoliated mesothelial cells. *Acta Cytologica*, **26**, 103–8.

Boon, M.E., Lindeman, J., Meeuwissen, A. and Otto, A.J. (1982b). Carcinoembryonic antigen in sputum cytology. *Acta Cytologica*, **26**, 389–94.

Boon, M.E., Wickel, A.F. and Davoren, R.A.M. (1983). Role of the air bubble in increasing cell recovery using cytospin I and II. *Acta Cytologica*, **27**, 699–702.

Boon, M.E., Veldhuizen, R.W., Ruinaard, C., Snieders, M.W. (1984a). Qualitative distinctive differences between the vacuoles of mesothelioma cells and of cells from metastatic carcinoma exfoliated in pleural fluid. *Acta Cytologica*, **28**, 443–8.

Boon, M.E., Veldhuizen, R.W. and Van Velzen, D. (1984b). Analysis of number, size and distribution patterns of lipid vacuoles. *Anal. Quant. Cytol.*, **6**, 221–5.

Bots, G.Th., Went, L.N. and Schaberg, A. (1964). Results of a sedimentation technique for cytology of cerebrospinal fluid. *Acta Cytologica*, **8**, 234.

Brown, A. and Scholtz, C.L. (1979). The quantitative histochemistry of RNA using gallocyanin. *Stain Technology*, **54** (2), 89–92.

Bulmer, D. (1959). Dimedone as an aldehyde blocking reagent to facilitate the histochemical demonstration of glycogen. *Stain Technology*, **34**, 95–8.

Busch, H. and Smetana, K. (1970). *The Nucleolus*. Academic Press, New York, pp. 549–75.

Busch, H., Lischwe, M.A., Michalek, J., Pui-Kwong Chan and Busch, R.K. (1982). Nucleolar proteins of special interest: silver staining proteins B23 and C23 and antigens of human tumour nucleoli. In: *The Nucleolus*. Society for Experimental Biology, Seminar Series 15 (Jordan, E.G. and Cullis, C.A., Eds). Cambridge University Press, Cambridge.

Caspersson, T., Auer, G., Fallenius, A. and Kudynowsky, J. (1983). Cytochemical changes in the nucleus during tumour development. *Histochem. J.*, **15**, 337–62.

Celalier, R.P. (1967). Tertiary (butanol–butyl alcohol) dehydrating of chromosome smears. *Stain Technology*, **31** (4), 155.

Chapman, C.B. and Whalen, E.J. (1947). The examination of serous fluids by the cell block technique. *N. Engl. J. Med.*, **237**, 215.

Chowdhury, T.R. and Chowdhury, J.R. (1981). Significance of the occurrence and distribution of glycogen in cervical cells exfoliated under different physiologic and pathologic conditions. *Acta Cytologica*, **25**, 557–65.

Clark, G. (1974). Comparison of various oxidants for alum hematoxylin. *Stain Technology*, **49** (4), 225–7.

Clark, G. (1975). Effects of additives on alum hematoxylin staining solutions. *Stain Technology*, **50** (2), 115–18.

Clark, G. (1979). Staining with chromoxane cyanine R. *Stain Technology*, **54** (6), 337–44.

Clark, G. (1981). *Staining Procedures*. Published for the Biological Staining Commission. Williams & Wilkins, Baltimore.

Cole, E.C. (1943). Studies on hematoxylin stains. *Stain Technology*, **18** (3), 125.

Comings, D.E. (1975). Mechanisms of chromosome banding IV. Optical studies of the Giemsa dyes. *Chromosoma*, **50**, 89–110.

Comings, D.E. (1978). Mechanisms of chromosome banding and implications for chromosome structure. *Ann. Rev. Genetics*, **12**,

25–46.

Comings, D.E. and Avelino, E. (1975). Mechanisms of chromosome banding. VII. Interaction of methylene blue with DNA and chromatin. *Chromosoma*, **51**, 365–79.

Couture, M.L., Freund, M. and Sedlis, A. (1979). The normal exfoliative cytology of menstrual blood. *Acta Cytologica*, **23**, 85–9.

Crabtree, W.N. and Murphy, W.M. (1980). The value of ethanol as a fixative in urinary cytology. *Acta Cytologica*, **24**, 452–5.

Das, D.K. and Chowdhury, J.R. (1981). The use of glycogen studies in the evaluation of treatment of carcinoma of the cervix uteri. *Acta Cytologica*, **25**, 566–71.

Davis, H.J. (1962). The irrigation smear. A cytological method for mass population screening by mail. *Am. J. Obst. Gynec.*, **84**, 1017.

Dean, W.W., Stasny, M. and Lubrano, G.J. (1977). The degradation of Romanowsky type blood stains in methanol. *Stain Technology*, **52** (1), 35–46.

Deitch, A.D. (1955). Microspectrometric study of the binding of the anionic dye, Naphthol Yellow S, by tissue sections and by purified proteins. *Lab. Invest.*, **4**, 324–51.

DeRobertis, E.D.P. and DeRobertis, E.M.F. (1980). *Cell and Molecular Biology*. Holt-Saunders International Editions.

Dhar, A.C. and Shah, C.K. (1982). Cytochemical method to localize acidic nuclear proteins. *Stain Technology*, **57** (3), 151–5.

Dracopoulou, I., Zambacos, J. and Lissaios, B. (1976). The value of rapid imprint smears in the surgery of skin cancer. *Acta Cytologica*, **20**, 553–5.

Drijver, J.S. and Boon, M.E. (1983*a*). Tertiary butanol as a substitute for xylene as a clearing agent. *Acta Cytologica.*, **27**, 210–11.

Drijver, J.S. and Boon, M.E. (1983*b*). Manipulating the Papanicolaou staining method: the role of activity in the EA counterstain. *Acta Cytologica*, **27**, 693–9.

Dunton, B.L. (1972). A simple counterstain method for cytological staining. *Acta Cytologica*, **16**, 361–2.

Ehrlich, P. (1956). *Collected Papers of Paul Ehrlich* (English edition). Pergamon, London.

Endo, Y., Morrii, T., Tamura, H. and Okuda, S. (1974). Cytodiagnosis of pancreatic malignant tumors by aspiration, under direct vision, using a duodenal fibroscope. *Gastroenterology*, 944–51.

Entschev, E.M. (1963). Eine neue Färbemethode in der Vaginalzytodiagnostik. *Zentralblatt für Gynäkologie*, **85**, 1008–1111.

Everett, M.M. and Miller, W.A. (1974). The role of phosphotungstic and phosphomolybdic acids in connective tissue studies. I. Histochemical studies. *Histochemical Journal*, **6**, 25–34.

Fernberg, M.R., Bhaskar, A.C. and Bourne, P. (1980). Differential diagnosis of malignant lymphomas by imprint cytology. *Acta Cytologica*, **24**, 16–25.

Fischer, S. (1978). Urine cytology with the filter imprint technique. In: *Symposium on Cytopathology*, Rio de Janeiro, Brazil.

Foot, N.C. (1938). Useful methods for the routine examination of brain tumors. *Am. J. Pathol.*, **XIV**, 245.

Foot, N.C. and Holmquist, N.D. (1958). Supravital staining of sediments of serous effusions; simple technique for rapid cytologic diagnosis. *Cancer*, **11**, 151–7.

Gabe, M. (1962). Résultats de l'histochemie des polysaccharides. (W. Graumann and K. Neumann, Eds). *Handbuch der Histochemie, Bd II, Polysacchariden*, 278–307.

Galbraith, W., Marshall, P.N., Lee, E.S. and Baccus, J.W. (1979). Studies on Papanicolaou staining: I. Visible-light spectra of stained cervical cells. *Anal. Quant. Cytol.*, **1** (3), 160–6.

Galbraith, W., Marshall, P.N. and Baccus, J.W. (1980). Microspectrophotometric studies of Romanowsky stained blood cells. I. Subtraction analysis of a standardized procedure. *J. Microsc.*, **119**, 313–29.

Gandolfi, A., Tedeschi, F. and Brizzi, R. (1983). The squash-smear technique in the diagnosis of spinal cord neurinomas. Report of three cases. *Acta Cytologica*, **27**, 273–6.

Gaub, J. (1981). Quantification of nuclear nonhistone proteins by Feulgen–Naphthol Yellow S cytophotometry. *Histochem. J.*, **13**, 717–22.

Gaub, J., Auer, G. and Zetterberg, A. (1975). Quantitative cytotechnical aspects of a combined Feulgen–Naphthol Yellow S staining procedure for the simultaneous determination of nuclear and cytoplasmic proteins and DNA in mammalian cells. *Exp. Cell Res.*, **92**, 323–32.

Giemsa, G. (1902*a*). Färbemethoden für Malariaparasiten. *Zentralblatt für Bakteriologie*, **I** (31), 429–30.

Giemsa, G. (1902*b*). Färbemethoden für Malariaparasiten. *Zentralblatt für Bakteriologie*, **I** (32), 307–13.

Giemsa, G. (1909). Uber die Färbung von Feucht-präparaten mit meiner Azurosinmethode. *Dtsch. med. Wochenschr.*, **35**, 1751.

Giemsa, G. (1910). Uber eine neue Schnellfärbung mit meiner Azur–Eosin–Lösung. *Münch. med. Wochenschr.*, **47**, 2476.

Gill, G.W. (1977). *Gill's modified OG-6 and EA for Papanicolaou staining*. Data sheet 196, Polysciences Inc., Warrington PA.

Gill, G.W., Frost, J.K. and Miller, K.A. (1974). A new formula for a half-oxidized hematoxylin solution that neither overstains nor requires differentiation. *Acta Cytologica*, **18**, 300–11.

Gilliland, J.W., Dean, W.W., Stasny, M. and Lubrano, G.J. (1979). Stabilised Romanowsky blood stain. *Stain Technology*, **54** (3), 141–50.

Godwin, J.T. (1976). Rapid cytologic diagnosis of surgical specimens. *Acta Cytologica*, **20**, 111–15.

Goldner, J. (1938). A modification of the Masson trichrome technique for routine laboratory purposes. *Am. J. Pathol.*, **14**, 237–43.

Gomori, G. (1936). Microtechnical demonstration of iron. *Am. J. Pathol.*, **12**, 655–63.

Gravlee, L.C. (1969). Jet irrigation method for the diagnosis of endometrial adenocarcinoma. Its principles and accuracy. *Obstet. Gynec.*, **34**, 168–73.

Gray, B. (1964). Sputum cytodiagnosis in bronchial carcinoma: A comparative study of two methods. *Lancet*, **2**, 549.

Gray, P., Pickle, F.M., Moser, M.D. and Hayweiser, L.J. (1956). Oxazine dyes I. Celestine blue B with iron as a nuclear stain. *Stain Technology*, **31** (4), 141–50.

Gurr, E. (1975). Some haematoxylin substitutes in biological microtechnique. *Laboratory Practice*, **24**, 155–8.

Hajdu, S.I. (1983). A note on the history of Carbowax in cytology. *Acta Cytologica*, **27**, 204–6.

Hance, R.T. and Green, F.J. (1959). Rapid ripening of hematoxylin solutions. *Stain Technology*, **34** (4), 237–8.

Hance, R.T. and Green, F.J. (1961). Behaviour of rapidly oxidized hematoxylin. *Stain Technology*, **36** (4), 253.

Hansch, C., Leo, A., Unger, S.H., Kim, H. and Lien, E.G. (1973). Aromatic substituent constants for structure–activity correlations. *J. Med. Chem.*, **16**, 1207–16.

Harris, M.J. and Keebler, C.M. (1976). Cytopreparatory techniques, In: *Compendium on Cytopreparatory Techniques* (Keebler, C.M., Reagan, J.W. and Wied, G.L., Eds). Tutorials on Cytology, Chicago, pp. 45–58.

Heidenhain, M. (1896). Noch einmal über die Darstellung der Zentralkörper durch Eisen-hämatoxylin nebst einigen allgemeinen Bemerk-ungen über die Hämatoxylinfarben. *Z. Wiss. Mikr.* 13, 186–99.

Hopwood, D. (1973). In: *Fixation in Histochemistry* (P.J. Stoward, ed.). Chapman & Hall, London.

Horobin, R.W. (1980). Structure–staining relationships in histochemistry and biological staining, *J. Microscopy*, **119** (3), 345–55, 357–72.

Horobin, R.W. (1982), *Histochemistry*. Butterworth, London.

Horobin, R.W. and Bennion P.J. (1973). The interrelation of the size and substantivity of dyes: the role of van der Waals attractions and hydrophobic bonding in biological staining. *Histochemie*, **33**, 191–204.

International Committee for Standardization in Haematology (1984). *ICSH Reference Method for Staining of Blood and Bone Marrow Films by Azure B and Eosin Y (Romanowsky–Giemsa Stain)*.

Isaacs, J.H. and Wilhoite, R.W. (1974). Aspiration cytology of the endometrium; office and hospital sampling procedures. *Am. J. Obstet. Gynecol.*, **118**, 679–84.

Iwama de Mattos, M.C.F. (1979). Cell-block preparation for cytodiagnosis of pulmonary paracoccidiodomycosis. *Chest*, **75**, 212.

Jimenez-Ayala, M., Vilaplana, E., DeBengoa, C.B., Zomeno, M., Moreno, S. and Granados, M. (1975). Endometrial and endocervical brushing techniques with a Medhosa cannula. *Acta Cytologica*, **19**, 557–63.

Johansen, P. and Thuen, D. (1981). Cytochemistry of imprints from breast biopsies. *Acta Cytologica*, **25**, 171–7.

Kölmel, H.W. (1977). A method for concentrating cerebrospinal fluid cells. *Acta Cytologica*, **21**, 154–7.

Koss, L.G. (1979). *Diagnostic Cytology*. J.B. Lippincott, Philadelphia.

Kramer, H. and Windrum, G.M. (1955). The

metachromatic staining reaction. *J. Histochem. Cytochem.*, **3**, 227.

Krentz, M.J. and Dyken, R.P. (1972). Cerebrospinal fluid cytomorphology: Sedimentation vs filtration. *Arch. Neurol.*, **26**, 253–7.

Kurnick, N.B. (1952). Histological staining with methyl green–pyronin. *Stain Technology*, **27** (5), 233.

Kwee, W.S., Veldhuizen, R.W., Alons, C.A., Morawetz, F. and Boon, M.E. (1982). Quantitative and qualitative differences between benign and malignant mesothelial cells in pleural fluid. *Acta Cytologica*, **26**, 401–6.

Lahiri, T. and Chowdhury, T.R. (1981). Lipid patterns in vaginal cells exfoliated from different physiologic conditions. *Acta Cytologica*, **25**, 572–6.

Lee, T.K. (1982). The value of imprint cytology in tumor diagnosis. A retrospective study of 522 cases in northern China. *Acta Cytologica*, **26**, 169–71.

Leeuwenhoek, A. van (1674, 1679, 1702). Letters to the Royal Society. *Phil. Trans. R. Soc. London*, **9**, 121; **12**, 1040; **22**, 552.

Lehninger, A.L. (1975). *Biochemistry*. Worth, New York.

Leif, R.C., Ingram, D., Clay, C., Bobbitt, D., Gaddis, R., Leif, S.B. and Nordquist, S. (1977). Optimization of the binding of dissociated exfoliated cervicovaginal cells to glass microscope slides. *J. Histochem. Cytochem.*, **25**, 538–43.

Liao, J.C., Ponzo, J.L. and Patel, C. (1981). Improved stability of methanolic Wright's stain with additive reagents. *Stain Technology*, **56** (4), 251–63.

Lillie, R.D. (1943a) Blood and malaria parasite staining with eosin azure methylene blue methods. *Am. J. Publ. Hlth*, **33**, 948–51.

Lillie, R.D. (1943b). A Giemsa stain of quite constant composition and performance made in the laboratory from Eosin and Methylene Blue. *Public Health Rep.*, **55**, 440–52.

Lillie, R.D. (1977). *H.J. Conn's Biological Stains*. Williams and Wilkins, Baltimore.

Lillie, R.D. (1978). Romanowsky–Malachowski stains, the so-called Romanowsky stain: Malachowski's 1891 use of alkali polychromed Methylene Blue for malaria plasmodia. *Stain Technology*, **53** (1), 23–8.

Lillie, R.D. and Donaldson, P.T. (1979). The wet Giemsa method for quick testing of variants in blood and malaria stains. *Stain Technology*, **54** (1), 47–8.

Lillie, R.D. and Fulmer, H.M. (1976). *Histopathologic Technique and Practical Histochemistry*. McGraw-Hill, New York.

Lillie, R.D., Pizzolato, P. and Donaldson, P.T. (1975a). Hematoxylin substitutes. Gallein as a biological stain. *Stain Technology*, **49** (6), 339–46.

Lillie, R.D., Pizzolato, P. and Donaldson, P.T. (1975b). Fluorone black and methyl fluorone black as metachrome iron alum mordant dyes. *Stain Technology*, **50**, 127–31.

Lillie, R.D., Pizzolato, P. and Donaldson, P.T. (1976a). Hematoxylin substitutes; a survey of mordant dyes tested and a consideration of their structure to performance as nuclear stains. *Stain Technology*, **51**, 25.

Lillie, R.D., Donaldson, P.T., Jirge, S.K. and Pizzolato, P. (1976b). Iron and aluminum lakes of gallo blue E as nuclear and metachromatic mucin stains. *Stain Technology*, **51**, 187–92.

Lindholm, K., Nordgren, H. and Akerman, M. (1979). Electron microscopy of fine needle aspiration biopsy from a malignant fibrous histiocytoma. Report of a case. *Acta Cytologica*, **25**, 399–401.

Llewellyn, B.D. (1974). Mordant blue 3. A readily available substitute for hematoxylin in the routine hematoxylin and eosin stain. *Stain Technology*, **49** (6), 347–9.

Llewellyn, B.D. (1978). Improved nuclear staining with mordant blue 3 as a hematoxylin substitute. *Stain Technology*, **53** (2), 73–7.

Löhr, W., Sohmer, I. and Wittekind, D. (1974). The azure dyes; their purification and physicochemical properties. I. Purification of azure A. *Stain Technology*, **49** (6), 359.

Löhr, W., Grubhofer, N., Sohmer, I. and Wittekind, D.H. (1975). The azure dyes: their purification and physicochemical properties. II. Purification of azure B. *Stain Technology*, **50** (3), 149–56.

Lopez Cardozo, P. (1977). *Atlas of Clinical Cytology*. Lippincott, Philadelphia.

Love, R., Takeda, M., Soriano, R.Z. and McCullough, L.B. (1973). The value of the internal structure of the nucleolus in the diagnosis of malignancy. *Acta Cytologica*, **17**, 310–15.

Luna, L.G. (1968). *Manual of Histologic Staining Methods of the Armed Forces Institute of Path-*

ology. McGraw-Hill, New York.

McCormick, W.F. and Coleman, S.A. (1962). A membrane filter technique for cytology of spinal fluid. *Am. J. Clin. Pathol.*, **38**, 191.

McLarty, J.W., Farley, M.L., Greenberg, S.D., Hurst, G.A. and Mabry, L.C. (1980). Statistical comparison of aerosol-induced and spontaneous sputum specimens in the Tyler Asbestos Workers' Program. *Acta Cytologica*, **24**, 460–5.

Mallory, F.B. (1900). A contribution to staining methods. *J. Exp. Med.*, **5**, 15–20.

Malmgren, R.A. (1965). Problems in techniques used in blood specimen preparation. *Acta Cytologica*, **9**, 97–9.

Marshall, P.N. and Horobin, R.W. (1973a). The mechanism of action of 'mordant' dyes—a study using preformed metal complexes. *Histochemie*, **35**, 361–71.

Marshall, P.N. and Horobin, R.W. (1973b). Measurements of the affinities of basic and 'mordant' dyes for various tissue substrates. *Histochemie*, **36**, 303–12.

Marshall, P.N. and Lewis, S.M. (1974). Batch variations in commercial dyes employed for Romanowsky-type staining: a thin layer chromatographic study. *Stain Technology*, **49** (6), 351–9.

Marshall, P.N. and Lewis, S.M. (1976). The purification of methylene blue and azure B by solvent extraction and crystallization. *Stain Technology*, **50**, 375–81.

Marshall, P.N., Galbraith, W. and Baccus, J.W. (1979). Studies on Papanicolaou staining. II. Quantitation of dye components bound to cervical cells. *Anal. Quant. Cytol.*, **1** (3), 169–78.

Marshall, P.N., Galbraith, W., Navarro, E.F. and Baccus, J.W. (1981). Microspectrophotometric studies of Romanowsky stained blood cells. II. Comparison of the performance of two standardized stains. *J. Microsc.*, **124** (2), 197–210.

Masson, P. (1929). Some histological methods. Trichrome stainings and their preliminary technique. *J. Tech. Meth. Bull. Int. Ass. Med. Mus.*, **12**, 75–90.

Mavec, P. (1967). Cytologic diagnosis from tumor tissue using the 'quick smear method' during operation. *Acta Cytologica*, **11**, 229–30.

Mayer, P. (1891). Uber das Färben mit Hämatoxylin. *Mitt. Zool. Stat. Neapel*, **X** (1), 170–82.

Meloan, S.N. and Puchtler, H. (1974). Iron alizarin blue S stain for nuclei. *Stain Technology*, **49** (5), 301.

Mendelson, D., Tas, J. and James, J. (1983). Cuprolinic Blue: a specific dye for single-stranded RNA in the presence of magnesium chloride. II. Practical applications for light microscopy. *Histochem. J.*, **15**, 1113–21.

Mennemeyer, R., Bartha, M. and Kidd, C.R. (1979). Diagnostic cytology and electron microscopy of fine needle aspirates of retroperitoneal lymph nodes in the diagnosis of metastatic pelvic neoplasms. *Acta Cytologica*, **23**, 370–3.

Möllendorff, W. von and M. (1924). Untersuchungen zur Theorie der Färbung fixierter Preparate. *Erg. Anat. u. Entw. Gesch.*, **25**, 1–65.

Mouriquand, J., Mouriquand, C., Petitpas, E., Marmet, J.L. and M.A. (1981). Differential nucleolar staining affinity with a modified Papanicolaou staining procedure. *Stain Technology*, **56** (4), 215–19.

Mundkur, B. and Greenwood, H. (1968). Amido Black 10B as a nucleolar stain for lymph nodes in Hodgkin's disease. *Acta Cytologica*, **12**, 218–25.

Murgatroyd, L.R. and Horobin, R.W. (1969). Specific staining of glycogen with Haematoxylin and certain anthraquinone dyes. *Stain Technology*, **44** (1), 59–62.

Nagasawa, T. and Nagasawa, S. (1983). Enrichment of malignant cells from pleural effusions by percoll density gradients. *Acta Cytologica*, **27**, 119–23.

Nedelkoff, B., Christopherson, W.M. and Harter, J.S. (1961). A method for demonstrating malignant cells in the blood. *Acta Cytologica*, **5**, 203–5.

Nielsen, M.L., Fischer, S., Högsborg, E. and Therckelsen, K. (1983). Adhesives for retaining prefixed urothelial cells after imprinting from cellulose filters. *Acta Cytologica*, **27**, 371–5.

Nielsen, N.H. (1972). Cytology by the filter imprint technique. *Acta Pathol. Microbiol. Scand.*, **80**, 47–53.

Nishimura, A., Den, N., Sato, H. and Takeda, B. (1973). Exfoliative cytology of the biliary tract with the use of saline irrigation under choledoscopic control. *Ann. Surg.*, **178**, 594–9.

Oud, P.S., Henderik, J.B.J., Huysmans, A.C.L.M., Pahlplatz, M.M.M., Hermkens, H.G., Tas, J., James, J. and Vooys, G.P. (1984). The use of Light Green and Orange II as quantitative protein stains, and their combination with the Feulgen method for the simultaneous determination of protein and DNA. *Histochemistry*, **80**, 149–57.

Ouelette, R.J. (1975). *Introductory Chemistry*. Harper & Row, New York.

Overton, E. (1890). *Zeit. wiss. Mikr.* (See Baker, 1970.)

Pak, H.Y., Yokota, S., Teplitz, R., Show, S.L. and Werner, J.L. (1981). Rapid staining techniques employed in fine needle aspirations of the lung. *Acta Cytologica*, **25**, 178–84.

Papanicolaou, G.N. (1928). New cancer diagnosis. In: *Proc. 3rd Race Betterment Conference*. Battle Creek, Race Betterment Fdn, p. 528.

Papanicolaou, G.N. (1933). The sexual cycle in the human female as revealed by vaginal smears. *Am. J. Anat.*, **52** (3), 519.

Papanicolaou, G.N. (1941). Some improved methods for staining vaginal smears. *J. Lab. Clin. Med.*, **26**, 1200–5.

Papanicolaou, G.N. (1942). A new procedure for staining vaginal smears. *Science*, **95**, 2469, 438–9.

Papanicolaou, G.N. (1954). *Atlas of Exfoliative Cytology*. Cambridge, Mass., Harvard University Press.

Papanicolaou, G.N. and Traut, H.F. (1941). The diagnostic value of vaginal smears in carcinoma of the uterus. *Am. J. Obst. Gynecol.*, **42** (2), 193.

Papanicolaou, G.N. and Traut, H.F. (1943). *Diagnosis of Uterine Cancer by the Vaginal Smear*. The Commonwealth Fund, New York.

Pauling, L., Corey, R.B. and Hayward, R. (1954). The structure of protein molecules. *Sci. Am.*, **191**, July, 51.

Pawlick, G.F. (1977). Pamihall stain. *Acta Cytologica*, **21**, 183.

Pearse, A.G.E. (1957). Solochrome dyes in histochemistry with particular reference to nuclear staining. *Acta Histochemica*, **4** (8), 95–101.

Pearse, A.G.E. (1980). *Histochemistry, Theoretical and Applied*, vol. I: *Preparative and Optical Technology*. Churchill Livingstone, Edinburgh.

Pearson, J.C., Kromhout, L. and King, E.B. (1981). Evaluation of collection and preservation techniques for urinary cytology. *Acta Cytologica*, **25**, 327–33.

Pickren, J.W. and Burke, E.M. (1963). Adjuvant cytology to frozen sections. *Acta Cytologica*, **7**, 164–7.

Pieslor, P.C., Oertel, Y.C. and Mendoza, M. (1979). The use of 2-molar urea as hemolyzing solution for cytologic smears. *Acta Cytologica*, **23**, 137–9.

Pool, E.H. and Dunlop, G.R. (1934). Cancer cells in the blood stream. *Am. J. Cancer*, **21**, 99–102.

Puchtler, H. and Isler, H. (1958). The effect of phosphomolybdic acid on the stainability of connective tissues by various dyes. *J. Histochem. Cytochem.*, **6**, 265–70.

Pundel, J.P. (1950). Les frottis vaginaux et cervicaux. Masson & Cie, Paris.

Rakoff, A.E. (1960). Review of techniques of vaginal smears. *Acta Cytologica*, **4**, 222.

Ranvier, L. (1875). *Traité Technique d'Histologie*. Libraire F. Savy, Paris.

Retief, A.E. and Rüchel, R. (1977). Histones removed by fixation: their role in the mechanism of chromosomal banding. *Exp. Cell Res.*, **106**, 233–7.

Reynaud, A.J. and King, E.B. (1967). A new filter for diagnostic cytology. *Acta Cytologica*, **11**, 289–94.

Rietveld, W.J. and Boon, M.E. (1981). Variations in size and glycogen content of exfoliated epithelial cells at different times of the day. *XVth Int. Conf. ISC*.

Romeis, B. (1968). *Mikroskopische Technik*. Oldenbourg Verlag, Munich.

Rubio, C.A. (1977). The false negative smear II: The trapping effect of collecting instruments. *Obstet. Gynecol.*, **49**, 576–80.

Rubio, C.A., Kock, Y. and Berglund, K. (1980). Studies of the distribution of abnormal cells in cytologic preparations. I. Making the smear with a wooden spatula. *Acta Cytologica*, **24**, 49–53.

Ruiter, D.J. and Boon, M.E. (1982). Atypical (reserve) cells in the cervical epithelium and their exfoliative pattern. *Acta Cytologica*, **25**, 292–8.

Ruiter, D.J., Mauw, B.J. and Beyer-Boon, M.E. (1979). Ultra-structure of normal epithelial cells in Papanicolaou-stained cervical smears. An application of a modified open-face embedding technique for transmission electron microscopy. *Acta Cytologica*, **23**, 507–15.

Saccomano, G., Saunders, R.P., Ellis, M., Archer, V.E., Wood, B.G. and Beckler, P.A. (1963). Concentration of carcinoma or atypical cells in sputum. *Acta Cytologica*, **7**, 305–10.

Sachdeva, R. and Kline, T.S. (1981). Aspiration biopsy cytology and special stains. *Acta Cytologica*, **25**, 678–83.

Sandritter, W., Diefenbach, H. and Krantz, F.

(1954). Uber die quantitatieve Bindung von Ribonukleinsäure mit Gallocyanin chromalum. *Experientia*, **10**, 210–15.

Sandritter, W., Kiefer, G. and Rick, W. (1966). *Introduction to Quantitative Cytochemistry* (Wied, G.L., ed.). Academic Press, New York.

Sayk, J. (1962). *Cytologie der Zerebrospinalflüssigkeit*. Fischer Verlag, Jena.

Schachter, A., Beckerman, A., Bahary, C. and Joel-Cohen, S.J. (1980). The value of cytology in the diagnosis of endometrial pathology. *Acta Cytologica*, **24**, 149–52.

Shorr, E. (1940*a*). I. A new technique for staining vaginal smears. *Science*, **91**, 321.

Shorr, E. (1940*b*). A new technique for staining vaginal smears II. *Science*, **91**, 579.

Shorr, E. (1942). A new technique for staining vaginal smears III. A single differential stain. *Science*, **94**, 545.

Shu, Y.J. (1983). Cytopathology of the esophagus. An overview of esophageal cytopathology in China. *Acta Cytologica*, **27**, 7–160.

Sills, B. (1953). Dry method for conserving and transporting cytological smears. *JAMA*, **151**, 230.

Singer, M. (1952). Factors which control the staining of tissue sections with acid and basic dyes. *Int. Rev. Cytol*, **1**, 211–55.

Smetana, K. and Busch, H. (1966). Studies on staining and localization of acidic nuclear proteins in the Walker 256 carcinosarcoma. *Cancer. Res.*, **26**, 334–7.

Smith, M.J., Kini, S.R. and Watson, E. (1980). Fine needle aspiration and endoscopic brush cytology. Comparison of direct smears and rinsing. *Acta Cytologica*, **24**, 456–9.

Spaander, P.J., Ruiter, D.J., Hermans, J., De Voogt, H.J., Brussee, J.A.M. and Boon, M.E. (1982). The implication of subjective recognition of malignant cells in aspirations of prostate cancer using cell image analysis. *Anal. Quant. Cytol.*, **4**, 123–7.

Spicer, S.S. (1961). Differentiation of nucleic acids by staining at controlled pH and by a Schiff-methylene blue sequence. *Stain Technology*, **36** (6), 337–40.

Spicer, S.S. and Lillie, R.D. (1961). Histochemical identification of basic proteins with Biebrich scarlet at alkaline pH. *Stain Technology*, **36** (6), 365–70.

Sumner, A.T. (1980). Dye binding mechanisms in G-banding of chromosomes. *J. Microsc.*, **119** (3), 397–406.

Sumner, A.T. and Evans, H.J. (1973). Mechanisms involved in the banding of chromosomes with Quinacrine and Giemsa. II. The interaction of the dyes with the chromosomal components. *Exp. Cell Res.*, **81**, 223–6.

Szczepanik, E. (1978). Zytologische Schnellfärbung in der gynäkologischen Sprechstunde. *Fortschr. Med.*, **96**, 804–6.

Taft, E.B. (1951). The problem of a standardized technique for methyl green–pyronin stain. *Stain Technology*, **26**, 205–12.

Takahashi, M. (1981). *Color Atlas of Cancer Cytology*. Thieme Verlag, Stuttgart.

Tas, J., van der Ploeg, M., Mitchell, J.P. and Cohn, N.S. (1980). Protein staining methods in quantitative cytochemistry. *J. Microsc.* **119**, 295–311.

Tas, J., Mendelsohn, D. and Van Noorden, C.J.F. (1983). Cuprolinic Blue: a specific dye for single-stranded RNA in the presence of magnesium chloride, I. Fundamental aspects. *Histochem. J.*, **15**, 801–14.

Thompson, P. (1982). Thin needle aspiration biopsy. *Acta Cytologica*, **26**, 262–3.

To, A., Dearneley, D.P., Ormerod, M.G., Canti, G. and Coleman, D.V. (1983). Indirect immunoalkaline phosphatase staining of cytologic smears of serous effusions for tumor marker studies. *Acta Cytologica*, **27**, 109–13.

Trott, P. (1983). Needle aspiration terminology. *Acta Cytologica*, **27**, 83.

Unna, P.G. (1902). Eine Modifikation der Pappenheimschen Färbung auf Granoplasma. *Monatsschr. prak. Dermat., Hamburg u. Leipzig*, **35**, 76–80.

Veeken, J.G.P.M. van der (1983). Biologische ritmen van glycogeen in vagina en wangepititheal. Leids Cytologisch Laboratorium.

Venkataraman, K. (1952). *The Chemistry of Synthetic Dyes*. 2 vols. Academic Press, New York.

Vickerstaff, T. (1950). *The Physical Chemistry of Dyeing*. Oliver & Boyd, London.

Wittekind, D.H. (1983). On the nature of Romanowsky–Giemsa staining and its significance for cytochemistry and histochemistry: an overall view. *Histochem. J.*, **15**, 1029–47.

Wittekind, D.H. and Hilgarth, M. (1979). Die einfache reproduzierbare Papanicolaou-Färbung. *Gebrutzhilfe und Frauenklinik*, **39** (11), 969–73.

Wittekind, D.H., Kretschmer, V. and Löhr, W. (1976). Kann Azur B–Eosin die May-Grünwald--Giemsa Färbung ersetzen? *Blut*, **32**, 71–8.

Wittekind, D.H., Kretschmer, V. and Sohmer, I. (1982). Azure B–eosin Y stain as the standard Romanowsky–Giemsa stain. *Brit. J. Haematol.*, **51**, 391–3.

Wood, M.L. and Green, A.G. (1958). Studies on textile dyes for biological staining. V. Pontacyl blue black SX, pontacyl violet 6R and luxol fast yellow TN. *Stain Technology*, **33**, 279–81.

Yam, L.T. and Janickla, A.J. (1983). A simple method of preparing smears from bloody effusions for cytodiagnosis. *Acta Cytologica*, **27**, 114–18.

Yasumatsu, H. (1977). Stain using celestine blue B as substitute nuclear stain in routine cytologic examinations. *Acta Cytologica*, **21**, 173–4.

Zajicek, J. (1974). *Aspiration Biopsy Cytology. Part I: Cytology of Supradiaphragmatic Organs.* (*Monographs in Clinical Cytology*, vol. 4). Karger, New York.

Zajicek, J. (1979). *Aspiration Biopsy Cytology. Part II: Cytology of Infradiaphragmatic Organs* (*Monographs in Clinical Cytology*, vol. 7). Karger, New York.

Zanker, V. (1981). Grundlagen der Farbstoff–Substrat Beziehungen. *Acta Histochemica, Suppl.* **XXIV**, S151–68.

Subject index

acetate buffer, 138
acetic acid, 114, 135, 147
    as acidifier, 72, 151, 152, 153,
        158, 160
    as fixative, 52, 53, 55, 56
acetone, 147
    as fixative, 52, 122, 138
    for dehydration, 65, 66
    in Wright's stain, 146
acid dyes, 42
Acid Fuchsin, 36, 60, 69, 71, 73,
    90, 91, *171*
acidophil cytoplasm, 100
acidophil granules, 100
Acridine Orange, 36, 84, *171*
Acridine Red–Malachite Green,
    65
additional nuclear staining, 63
additive fixatives, 56
adenine, 5, 6, 40
adhesives, 116
air-dry methods, 119
air-drying, 53, 54, 55, 78, 80, 85,
    94, 98, 101, 117, 123, 129,
    138, 140, 142, 143, 146, 147
    and fat staining, 82
    effects, 55
albumin, 55
    for the prevention of cell
        explosion during air-
        drying, 55
Alcian Blue, 36, 81, *172*
    to stain mucin, 14, 141
Alcian Blue method, 82, 141
alcohol (*see also* ethyl alcohol), 22,
    121
    as differentiator, 139

as fixative, 130, 131, 132, 133,
    138, 139
as solvent, 157, 158, 159, 165,
    167
for destaining, 148
aldose, 13
Alizarin Blue, 36, *172*
    as natural dye, 41
Alizarin Cyanin BB, 64
aluminium alum, 62, 63, 151, 152
aluminium sulphate, 141, 143, 153
Amido Black 10B, 37, 64, 85, 137,
    173
amino acid
    as basic units of peptides, 8, 9
    in histones, 7
ammonium alum, 151, 153
ammonium carminate, 90, 91
ammonium molybdate, 85, 137,
    157
ammonium oxalate, 166
ammonium sulphate, 142, 147
amphoteric dye, 35, 38
amphoteric Haematein, 61
amphoteric Light Green, 73
amylase treatment, 80, 140
Aniline Blue, 36, 69, 72, *173*
anionic dye, 35
Anthocyanins as natural dyes, 41
Apathy's gum syrup, 124
arginine, 7, 9
artifacts, 59
ascites, 146
    aspiration fluid, 146
asparagine as hydrophilic agent in
    proteins, 13
aspartic acid, 9, 49

aspiration, 94
    cytology, 28, 105
Auer's rods, 145
autolysis, 51
auxochrome, 34
Ayre spatula, 102
Azocarmine, 60
Azure A, 37, 64, 65, 88, 93, 95, 96,
    138, 166, *174*
Azure B (synonym of Azure I and
    Methylene Azure), 37, 88,
    94, 95, 96, 97, 98, 99, 100,
    165, 166, *174*
Azure C, 37, 96, *175*
Azure I, *see* Azure B
Azure II, 93, 165
Azures as metachromatic dyes, 47

bacteria, staining of, 147, 167
bacteriostatic agents, 103, 116
Balayette brush, 102
balloon techniques, 107
Barr body, 3, 4, 138, 139, 176
basic dyes, 42
Basic Fuchsin, 36, *175*
    in Schiff's reagent, 38, 157
basic proteins, 68, 135
Best's carmine, 14, 41, 162
    to stain glycogen, 14, 139, 177
Biebrich Scarlet, 37, 68, 70, 79,
    139, *176*
    as protein dye, 50
    in Shorr's stain, 79, 160
biopsies, 116
Bismarck Brown, 37, 68, 69, 71,
    73, 88, *176*
blebs, 81

* Numbers in *italics* indicate the page on which the structural formula appears.

Part Two

1.1

1.2

1.3

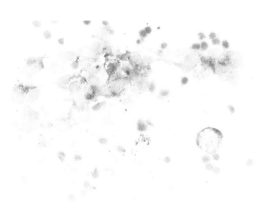

1.4

PLATE 1 Microscopy without staining. **1.1**, Using polarised light. Cholesterol crystals in aspiration of tumour on head. Tissue diagnosis: atheroma. **1.2**, Using polarised light, with quartz red I filter, unstained. Starch crystals. **1.3**, Brown colour of melanin in aspiration of lymph node metastasis of malignant melanoma. The slide is unstained. **1.4**, Same case as 1.3. The melanin is stained with the modified Schmorl method (number 38).

2.1

2.2

2.3

2.4

PLATE 2   Components of the nuclear and cytoplasmic stains in the Papanicolaou method. **2.1**, Haematoxylin only (method 3 without acid bath). Vaginal smear with intermediate cells. Note that the cytoplasm also stains blue. **2.2**, Orange G, pH 2.5. Anucleated highly keratinised squames from the foot. Note that the colour is yellow. **2.3**, Leiden EA (number 14). No Haematoxylin. Note turquoise colour of cytoplasm and nucleus. Vaginal smear. **2.4**, Eosin Y, 0.5% with 0.2 g PTA. Note pink colour of these intermediate cells, and compare with plate 2.3.

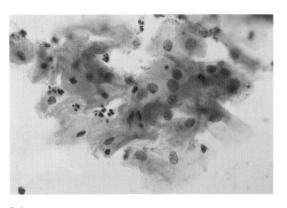

3.1

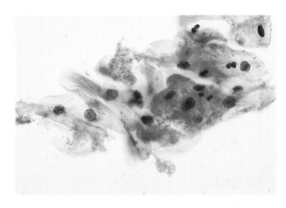

3.2

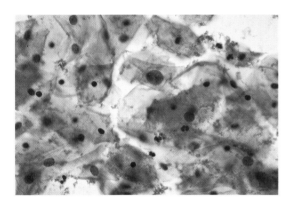

3.3

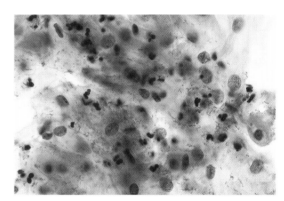

3.4

PLATE 3 Comparison of EA with different pH values (omitting Bismarck Brown). **3.1**, Smear from slight dysplasia. EA has pH of 4.6. No keratinised cells present. Cytoplasm stains turquoise. **3.2**, Same case as 3.1. EA has pH 6.5. Cytoplasm stains blue. **3.3**, Smear from condylomatous dysplasia (no koilocytotic cells in this photographic field). Keratinised cells stain red, non-keratinised turquoise. Staining method 2. **3.4**, Same case as 3.3. The keratinised cells stain the same; the non-keratinised cells stain blue instead of turquoise. Staining method 3.

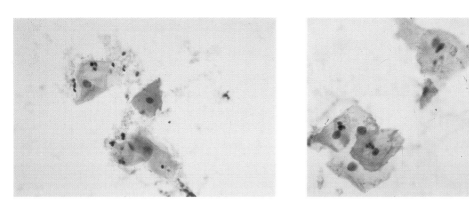

4.1   **4.2**

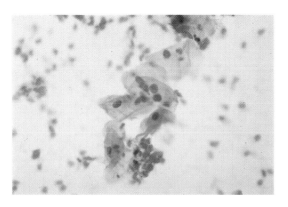

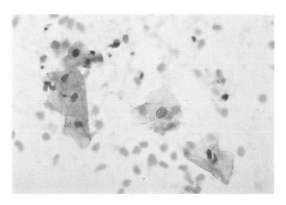

4.3   **4.4**

PLATE 4   Order of staining dishes in Papanicolaou methods including Orange G (pH 2.5). **4.1**, Staining method 1, pH of EA 4.6. Normal sequence, first Orange G, followed by EA. Superficial cells stain orange, intermediate cells stain turquoise. **4.2**, Same baths as in 4.1 but in reverse order (first EA, followed by Orange G). Here the orange G 'wins' over the Eosin, and thus the colour of the superficial cells is yellower. The intermediate cells are slightly destained. **4.3**, Staining method 1, with laboratory-made EA, pH 6.5. Normal sequence, first Orange G, followed by EA. Here, the Eosin has replaced the Orange G in the superficial cells, colouring them red. Note orange colour of erythrocytes. **4.4**, Same staining method as in 4.3, but reversed order (first EA, followed by Orange G). Here the Orange G has replaced the Eosin, staining erythrocytes and superficial cells orange–yellow. The intermediate cells are slightly destained.

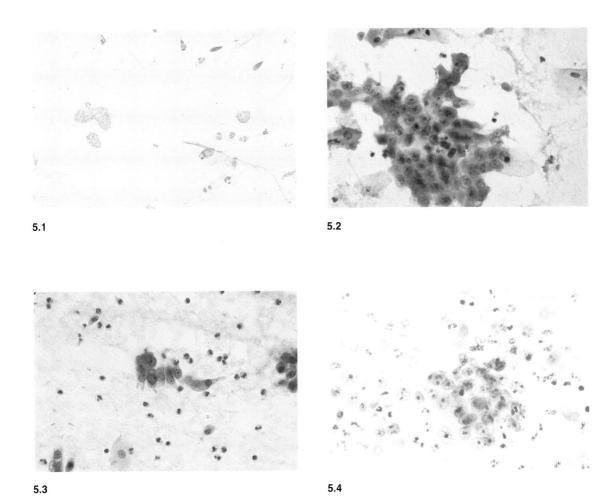

5.1

5.2

5.3

5.4

PLATE 5   Colour of nucleoli in the Papanicolaou methods. **5.1**, Method 2, using EA, pH 3.5. Colour of nucleoli is *green*, due to staining with Light Green. Undifferentiated carcinoma. **5.2**, Method 1, using EA, pH 6.5. Colour of nucleoli is *orange* due to staining with Orange G. Benign endocervical cells. **5.3**, Method 3. Colour of nucleoli is *red*, due to staining with Eosin. Cells from ovarian adenocarcinoma in vaginal smear. **5.4**, Method 3. Colour of nucleoli is *blue*, due to staining with Haematoxylin. Cells from endometrial adenocarcinoma.

6.1

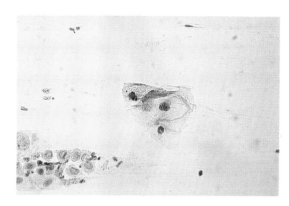

6.2

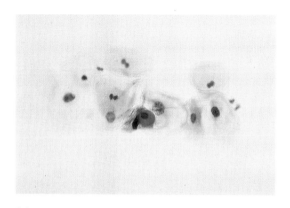

6.3

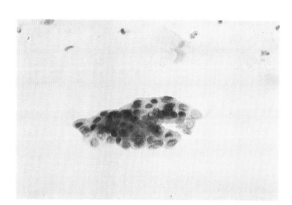

6.4

PLATE 6   Staining patterns of OG in cervical smears using Papanicolaou method 1. **6.1**, Koilocytotic cells. The rim of the koilos, containing dense cytoplasm, stains orange–yellow with the OG. **6.2**, Same cells as in 6.1, visualised with the Nomarski method. Note that the rim of the koilos is the thickest part of the cell. **6.3**, Case of condylomatous dysplasia. The small parakeratotic cells and some of the superficial cells are highly keratinised and stain orange–yellow. **6.4**, Group of endocervical cells. Note that the inner part of these cells (*without* signs of keratinisation) stains orange–yellow.

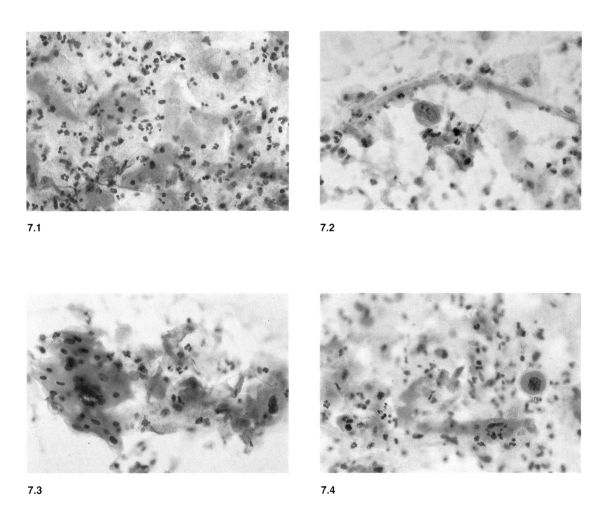

7.1

7.2

7.3

7.4

PLATE 7   Different steps of keratinisation visualised in the Orange G–Papanicolaou method (1). **7.1–7.4**, Smears from a verruceous carcinoma of the cervix. The most highly keratinised cells stain yellow, with decreasing keratinisation orange–red–blue. Even when the malignant keratinised cells have lost their nuclei, or contain small pyknotic nuclei, they can still be recognised as originating from the carcinoma by their abnormal shape.

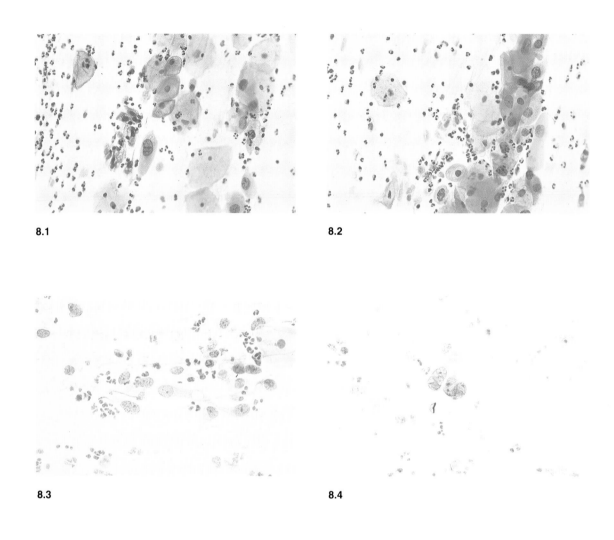

8.1

8.2

8.3

8.4

PLATE 8 Using the Papanicolaou method (3) on air-dried smears incubated overnight with formol–saline (van der Griendt's post–fixation method). **8.1 and 2**, Slight dysplasia, note good stainability of the nuclei with Haematoxylin. **8.3 and 4**, Adenocarcinoma in situ of endocervix. Note red nucleoli.

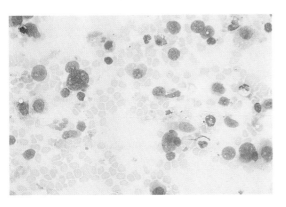

9.1

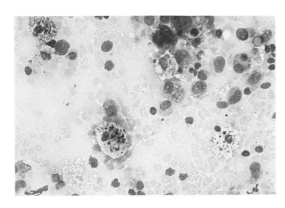

9.2

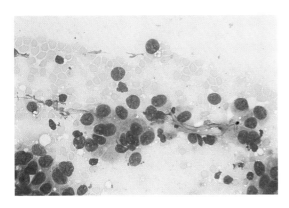

9.3

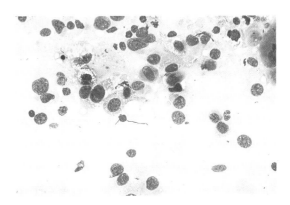

9.4

PLATE 9 Length of staining times in the Romanowsky–Giemsa method. Adenocarcinoma of the lung metastasis in lymph node. Method 35 (changing staining times), pH Giemsa 6.4. **9.1**, 15 min: poorly developed Romanowsky effect, bluish–purple nuclei. Note orange–pink staining of erythrocytes. **9.2**, 25 min: better developed Romanowsky effect. **9.3**, 35 min: well developed Romanowsky effect. **9.4**, 45 min: well developed Romanowsky effect. Nuclear pattern more 'open', and therefore nucleoli more visible.

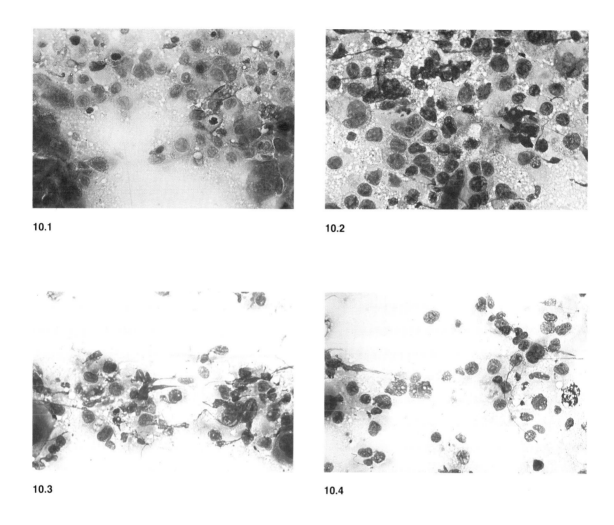

10.1

10.2

10.3

10.4

PLATE 10 Effect of pH in the Romanowsky–Giemsa method. Adenosquamous carcinoma of the lung, method 35 (changing pH of Giemsa). **10.1**, pH Giemsa 5.5. No Romanowsky effect, blue nuclei. **10.2**, pH Giemsa 6.4. Good Romanowsky effect. Colour of mucoid material pink, of (few) erythrocytes orange–red. **10.3**, pH Giemsa 7.0. Good Romanowsky effect. Colour of mucoid material grey–pink, of erythrocytes (not in microscopic field) grey. **10.4**, pH Giemsa 8.5. Good Romanowsky effect. Colour of mucoid material blue, of erythrocytes (not in microscopic field) green–blue.

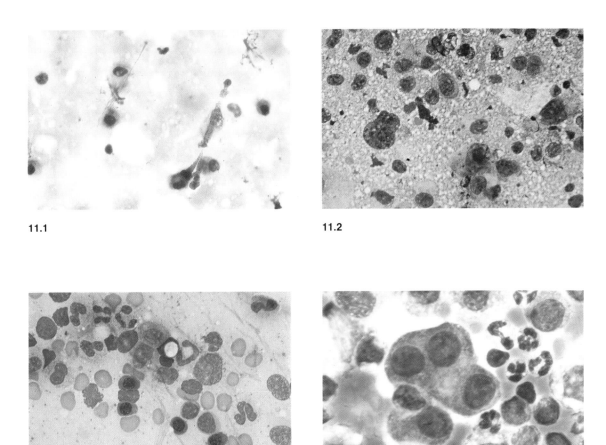

11.1

11.2

11.3

11.4

PLATE 11 Eosin staining in the Romanowsky–Giemsa method. **11.1**, Pleural fluid; asbestos body. The protein coat stains orange with Eosin. Method 35. **11.2**, Lung brush, poorly differentiated adenocarcinoma. Method 35, pH Giemsa 6.4. Erythrocytes stain orange with Eosin. **11.3**, Bone marrow, stained with Difquick method (commercial). Granulae stain orange with Eosin. Note grey colour of erythrocytes. **11.4**, Pleural fluid, mesothelioma. Method 34. Eosin plays a role in the pinkish staining of collagen in the centre of the mesothelioma cell grouping.

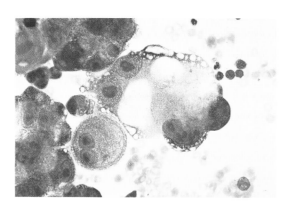

12.1

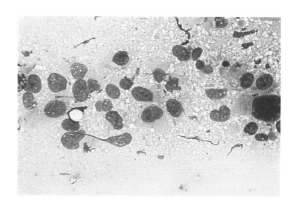

12.2

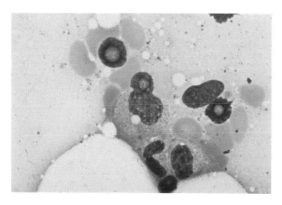

12.3

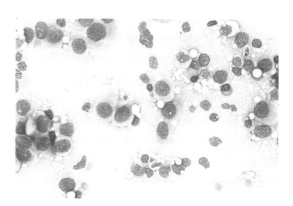

12.4

PLATE 12   Nucleoli in the Romanowsky–Giemsa method. **12.1**, Pleural fluid, metastasis of ovarian adenocarcinoma, method 34. In the well spread cells the blue nucleoli are clearly visible in the purple 'open' chromatin. In the thick cell grouping the nuclei are blue, and thus the blue nucleoli are invisible. **12.2**, Lung brush, adenocarcinoma, method 35. The purple chromatin is more dense than in 12.1 and therefore the nucleoli are not clearly visible. **12.3**, Bone marrow, immunocytoma, Difquick (commercial). Fat droplets on three of the nuclei, obscuring nucleoli. Blue nucleoli in the malignant immunocyte. **12.4**, Aspirate of carcinoma of the breast, staining method 33. Nucleoli difficult to distinguish in dense, purple chromatin.

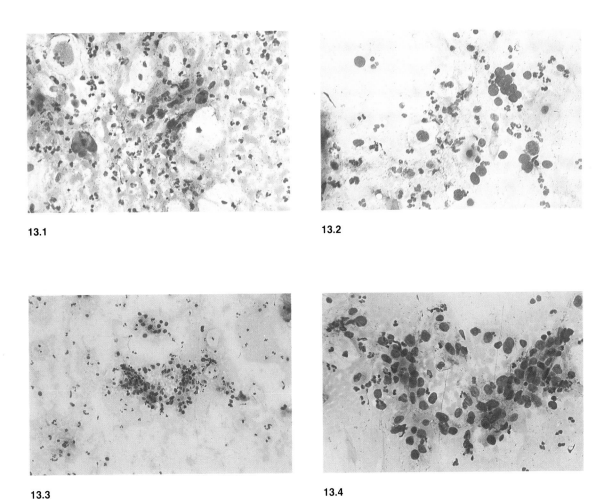

13.1

13.2

13.3

13.4

PLATE 13 Romanowsky–Giemsa method on air-dried cervical smears. All smears are stained with method 36. **13.1,** The spindle cells originating from squamous carcinoma of the cervix have azure-blue cytoplasm and dense purple nuclei. Note the orange colour of erythrocytes. **13.2,** The cells from undifferentiated carcinoma in situ of the cervix have large purple nuclei and no cytoplasm. **13.3,** Screening magnification of 13.4. **13.4,** Large cell non-keratinised carcinoma in situ of the cervix. The purple chromatin of the malignant cells shows a malignant pattern; the cytoplasm is pale blue.

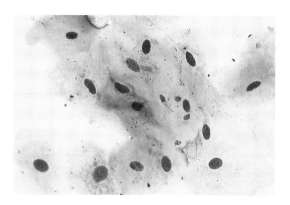

14.1

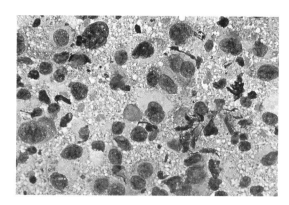

14.2

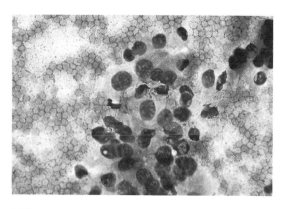

14.3

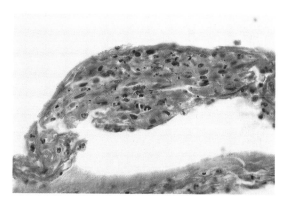

14.4

PLATE 14   Squamous differentiation in the Romanowsky–Giemsa method. **14.1**, Buccal smear, staining method 36, pH 7.0. Some cells stain azure-blue but most stain blue. **14.2**, Lung brush, adenosquamous carcinoma, method 35 with pH Giemsa of 6.4. One malignant squamous cell has azure-blue cytoplasm. **14.3**, Lung brush, moderately differentiated squamous cell carcinoma of the lung, staining method 35. The cytoplasm of the malignant cells stains pale blue. **14.4**, Blocked sputum, formalin fixed, staining for 60 min in 2:8 Giemsa. The cytoplasm of the malignant squamous cells is red.

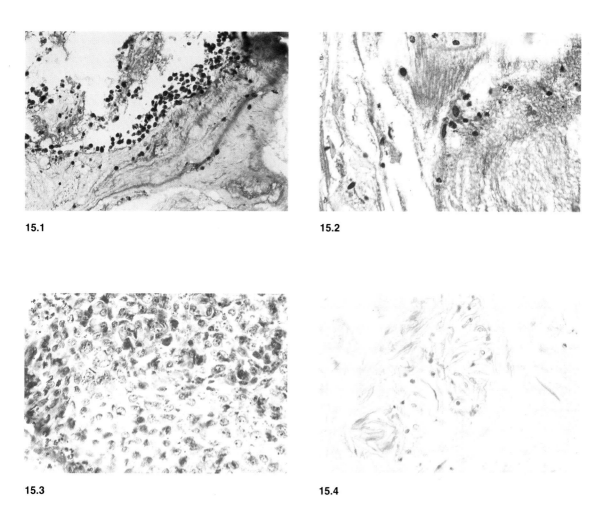

15.1

15.2

15.3

15.4

PLATE 15 Romanowsky–Giemsa method on cell blocks. The sections are deparaffinised in distilled water for 60 min in 2:8 Giemsa solution, dipped in 98% acetic acid, followed by 96% ethyl alcohol and dehydrated in isopropanol. **15.1**, Alcohol-fixed sputum, Giemsa pH 6.8. The nuclei of the oat cells are purple. **15.2**, Alcohol-fixed sputum, Giemsa pH 6.8. The nuclei of the malignant squamous cells are purple, the cytoplasm is red. **15.3**, Formalin-fixed sputum, Giemsa pH 6.8. A group of adenosquamous carcinoma cells. Due to the influence of formalin there is no Romanowsky effect, thus the nuclei stain blue. **15.4**, Alcohol-fixed sputum, Giemsa pH 4.8. Due to the low pH of the Giemsa there is no Romanowsky effect, thus the nuclei stain blue.

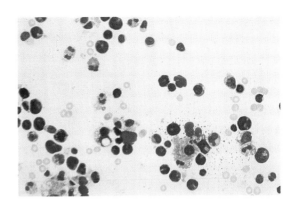

16.1

16.2

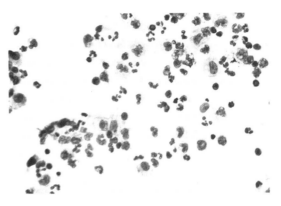

16.3

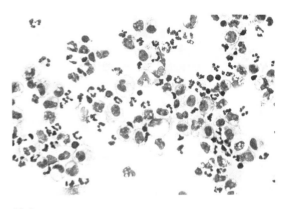

16.4

PLATE 16 Effects of protein in the air-drying process for the Romanowsky–Giemsa method, and in wet-fixation. **16.1**, T cell lymphoma in pleural fluid, high protein content. Smear from centrifuge sediment. Note that, due to the high protein content, the cells are not well spread, and therefore nuclear cleavage is not visible. **16.2**, Same case as 16.1, Cytospin slide. The protein is absorbed by the filter paper, and thus the malignant lymphoma cells do not dry in a high protein environment as in 16.1. Now, the nuclear cleavage of the malignant cells is clearly visible. **16.3**, Pleural fluid, inflammation. The Cytospin slide is wet-fixed with spray fixative. The protein in the background forms a network and therefore cytoplasm of the histiocytes cannot be delineated. **16.4**, Same case as 16.3, Cytospin slide, air-dried. Method 35. There is no interfering protein network and the histiocytes are well spread.

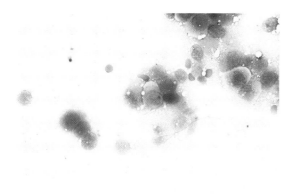

17.1

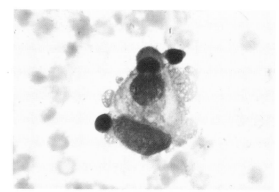

17.2

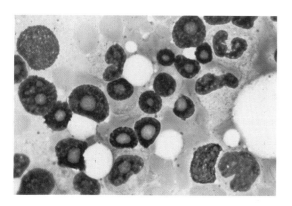

17.3

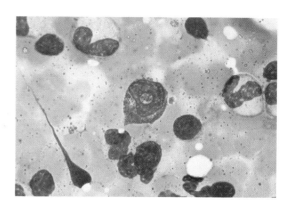

17.4

PLATE 17 Pleasant and unpleasant effects of air-drying for the Romanowsky–Giemsa method. **17.1**, Pleural fluid, metastasis of adenocarcinoma of the breast. The sediment in the centrifuge tube was not dry enough, and thus the cell spread contained too much fluid of a low protein content. The cells look 'watery' and are not suited for diagnostic work. **17.2**, Pleural fluid, mesothelioma. The laminar part of the cytoplasm is spread over the slide, forming 'blebs'. These stain positive with the PAS method, and are characteristic for mesothelial cells. **17.3**, Bone marrow, immunocytoma, Difquick (commercial). Fat globules on top of nuclei forming 'holes'. **17.4**, Bone marrow, staining method 34. The central cell has a true hole in the nucleus (note granulae in it).

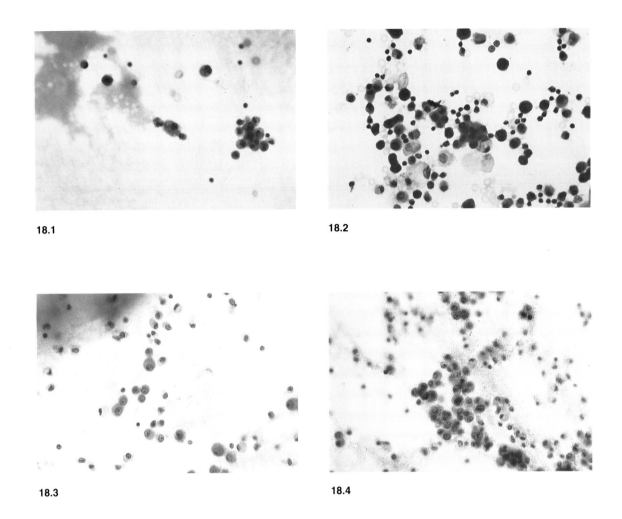

18.1

18.2

18.3

18.4

PLATE 18 Fibrin and protein in pleural fluid. **18.1**, Pleural fluid with high protein content, wet-fixed, staining method 3. **18.2**, Same case, air-dried smear from centrifuge sediment. Note that the cells are not well spread, and therefore there is no Romanowsky effect. Everything stains blue. **18.3**, Pleural fluid, fibrin in background. Cells better for diagnostic work than in 18.1. **18.4**, Same case as 18.3. Nomarsky. Note granular appearance of fibrin.

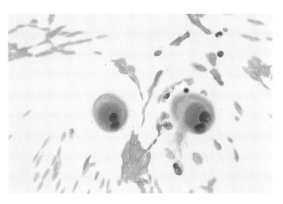

19.1

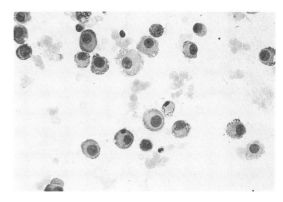

19.2

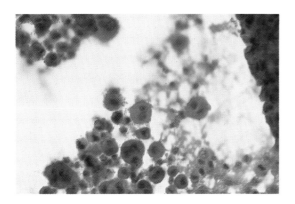

19.3

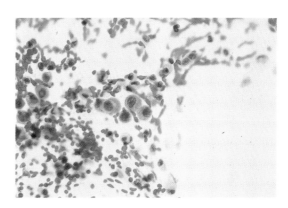

19.4

PLATE 19 Staining characteristics of mesothelial cells. **19.1**, Pleural fluid, wet-fixed. Staining method 4. Malignant mesothelioma. Note the 'two-tone' staining of the cytoplasm, indicating mesothelial differentiation. **19.2**, Same case as 19.1, air-dried. Giemsa staining method 34. No two-tone staining, but the laminar part of the cytoplasm stains darker blue (see plate 17). This is *not* the same area as the blue staining part of the two-tone cytoplasm in 19.1. In the latter cells, there is only minimal spreading of the laminar part of the cytoplasm. **19.3**, Same case as 19.1, same staining. This is a thicker part of the smear; here the malignant mesothelial cells stain orange-red, and are not recognisable as such by their staining pattern. **19.4**, Same case as 19.1. Staining Method 4. Smaller mesothelial cells with blue staining cytoplasm and no two-tone effect.

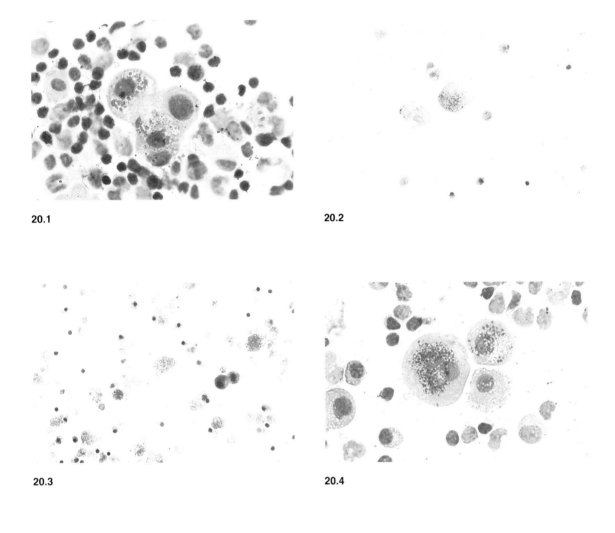

20.1

20.2

20.3

20.4

PLATE 20   Staining patterns, ORO stain (Method 27). **20.1**, Pleural fluid, malignant mesothelioma. Note characteristic peri- and supranuclear distribution of small ORO-positive vacuoles. **20.2**, Pleural fluid, metastasis of adenocarcinoma of the breast. Adenocarcinoma cell with non-perinuclear fat. **20.3**, Pleural fluid, malignant mesothelioma. Many histiocytes with non-perinuclear fat. **20.4**, Pleural fluid, same case as 19.1. Note that some of the peri- and supranuclear vacuoles have lost their fat content, leaving white 'holes' in the nucleus and cytoplasm.

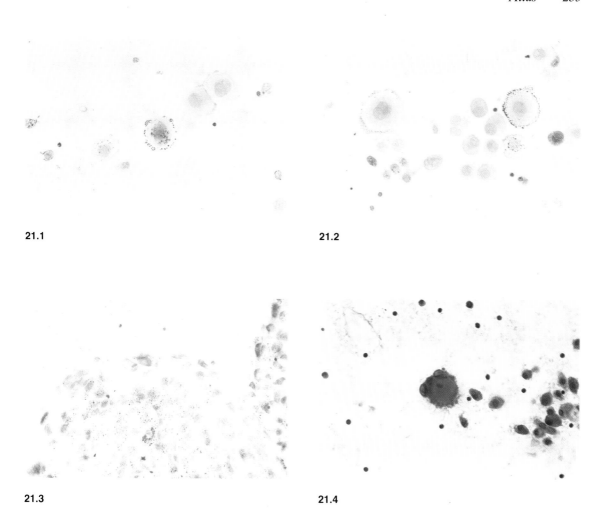

21.1

21.2

21.3

21.4

PLATE 21 Staining patterns, PAS stain (Method 23). **21.1 and 21.2**, Pleural fluid, malignant mesothelioma. Note the positive staining of the 'blebs' (see plates 17 and 19), characteristic for mesothelial cells. In addition, in 21.1 there is some central positive staining. **21.3**, Pleural fluid, metastasis of serous papillary carcinoma of the ovary. The positive dots are evenly divided over the cytoplasm. **21.4**, Ascites, metastasis of signet-ring carcinoma of the stomach. The PAS-positive large vacuole of the malignant signet ring compresses the nucleus, which therefore cannot be studied. The positive diagnosis of this fluid was based on the presence of these cells, which could not be recognised as malignant in the Papanicolaou and the Giemsa slides.

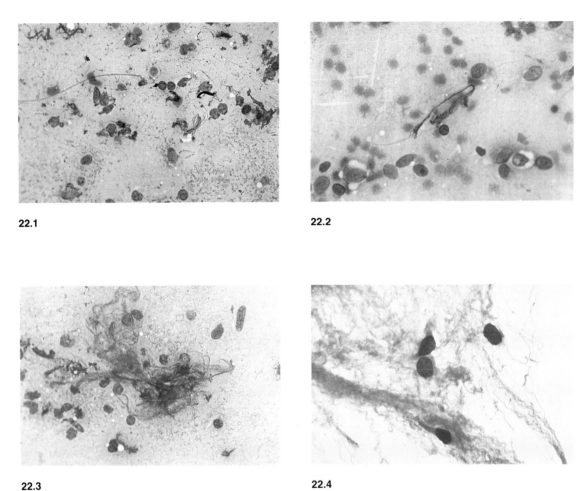

22.1

22.2

22.3

22.4

PLATE 22   Staining reticulum fibres (Method 28). **22.1**, Aspirate, fibrosarcoma. The fibres (thick bundles) are black (Method 28). **22.2**, Same case as 22.1, Giemsa method 35. Indication of a thick bundle of fibres. **22.3**, Aspirate, fibromyxosarcoma. Network of black fibres (Method 28). **22.4**, Same case as 22.3. Myxomatous mass with feathery aspect, containing malignant cells. Giemsa method 36.

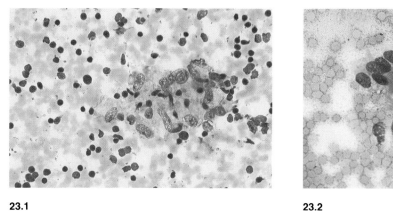

23.1

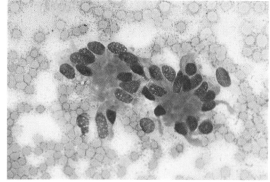

23.2

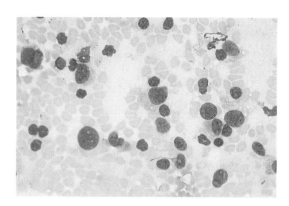

23.3

23.4

PLATE 23 Cell arrangements. **23.1**, Lymph node aspirate method 34, sarcoidosis. The epithelioid cells are arranged in dense clusters, and the lymphocytes have a 'starry sky' distribution. **23.2**, Bronchial brush, method 36. Benign bronchial cells. The nuclei are arranged on the outer border of the cell cluster which displays good cohesion. Note purplish-red brush border of the cells visible in the centre of the cell grouping. **23.3**, Aspirate, method 35, fibromyxosarcoma. The nuclei of the malignant cells are scattered over the microscopic field without a definite pattern, but seem to be attached to each other by the fibres. **23.4**, Aspirate, method 35, pH 6.4. Poorly differentiated adenocarcinoma of the lung. The nuclei are scattered over the microscopic field, sometimes in very loose small clusters. This pattern is different from the 'starry sky' appearance of the lymphocytes in 23.1.

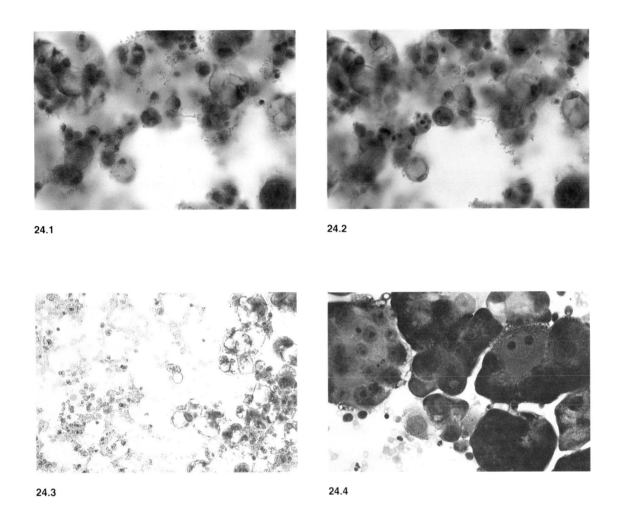

24.1

24.2

24.3

24.4

PLATE 24   Three-dimensionality in the Papanicolaou and the Romanowsky–Giemsa methods. **24.1 and 24.2**, Pleural fluid, metastasis of ovarian adenocarcinoma, alcohol fixed. Staining method 4. Two focusing planes of the same microscopic field. Note that different parts of the cells are in focus. **24.3**, Same preparation as 24.1. Nomarsky. Compare three-dimensional cell groupings of malignant cells (right) with flat histiocytes (left). **24.4**, Same case, air-dried, Giemsa method 34. Note absence of Romanowsky effect in thick cell groupings.